Generations Gardening Together

Sourcebook
for Intergenerational
Therapeutic Horticulture

FOOD PRODUCTS PRESS®
Crop Science
Amarjit S. Basra, PhD
Editor in Chief

The Lowland Maya Area: Three Millennia at the Human-Wildland Interface edited by A. Gómez-Pompa, M. F. Allen, S. Fedick, and J. J. Jiménez-Osornio

Biodiversity and Pest Management in Agroecosystems, Second Edition by Miguel A. Altieri and Clara I. Nicholls

Plant-Derived Antimycotics: Current Trends and Future Prospects edited by Mahendra Rai and Donatella Mares

Concise Encyclopedia of Temperate Tree Fruit edited by Tara Auxt Baugher and Suman Singha

Landscape Agroecology by Paul A. Wojtkowski

Concise Encyclopedia of Plant Pathology by P. Vidhyasekaran

Molecular Genetics and Breeding of Forest Trees edited by Sandeep Kumar and Matthias Fladung

Testing of Genetically Modified Organisms in Foods edited by Farid E. Ahmed

Fungal Disease Resistance in Plants: Biochemistry, Molecular Biology, and Genetic Engineering edited by Zamir K. Punja

Plant Functional Genomics edited by Dario Leister

Immunology in Plant Health and Its Impact on Food Safety by P. Narayanasamy

Abiotic Stresses: Plant Resistance Through Breeding and Molecular Approaches edited by M. Ashraf and P. J. C. Harris

Teaching in the Sciences: Learner-Centered Approaches edited by Catherine McLoughlin and Acram Taji

Handbook of Industrial Crops edited by V. L. Chopra and K. V. Peter

Durum Wheat Breeding: Current Approaches and Future Strategies edited by Conxita Royo, Miloudi M. Nachit, Natale Di Fonzo, José Luis Araus, Wolfgang H. Pfeiffer, and Gustavo A. Slafer

Handbook of Statistics for Teaching and Research in Plant and Crop Science by Usha Rani Palaniswamy and Kodiveri Muniyappa Palaniswamy

Handbook of Microbial Fertilizers edited by M. K. Rai

Eating and Healing: Traditional Food As Medicine edited by Andrea Pieroni and Lisa Leimar Price

Physiology of Crop Production by N. K. Fageria, V. C. Baligar, and R. B. Clark

Plant Conservation Genetics edited by Robert J. Henry

Introduction to Fruit Crops by Mark Rieger

Generations Gardening Together: Sourcebook for Intergenerational Therapeutic Horticulture by Jean M. Larson and Mary Hockenberry Meyer

Agriculture Sustainability: Principles, Processes, and Prospects by Saroja Raman

Introduction to Agroecology: Principles and Practice by Paul A. Wojtkowski

Handbook of Molecular Technologies in Crop Disease Management by P. Vidhyasekaran

Handbook of Precision Agriculture: Principles and Applications edited by Ancha Srinivasan

Dictionary of Plant Tissue Culture by Alan C. Cassells and Peter B. Gahan

Handbook of Potato Production, Improvement, and Postharvest Management edited by Jai Gopal and S. M. Paul Khurana

Generations Gardening Together
Sourcebook for Intergenerational Therapeutic Horticulture

Jean M. Larson, MA, CTRS, HTR
Mary Hockenberry Meyer, PhD

Routledge
Taylor & Francis Group

LONDON AND NEW YORK

First published 2006 by Haworth Press, Inc

Published 2018 by Routledge
52 Vanderbilt Avenue, New York, NY 10017
2 Park Square Milton Park, Abingdon Oxon OX14 4RN

Routledge is an imprint of the Taylor & Francis Group, an informa business

Cover design by Kerry E. Mack.
TR: 9.21.06

Library of Congress Cataloging-in-Publication Data

Larson, Jean M.
    Generations gardening together : sourcebook for intergenerational therapeutic horticulture / Jean M. Larson, Mary Hockenberry Meyer.
        p. cm.
    Includes bibliographical references and index.
    ISBN-13: 978-1-56022-319-1 (hard : alk. paper)
    ISBN-10: 1-56022-319-7 (hard : alk. paper)
    ISBN-13: 978-1-56022-320-7 (soft : alk. paper)
    ISBN-10: 1-56022-320-0 (soft : alk. paper)
    1. Occupational therapy for older people. 2. Gardening—Therapeutic use. I. Meyer, Mary Hockenberry. II. Title.
RC953.8.022L37 2006
615.8'515—dc22

                                                                        2005020814

ISBN 13: 9781560223191 (hbk)
ISBN 13: 9781560223207 (pbk)

Dedicated to Maggie Kuhn,
founder of the Gray Panthers
(1905-1995)

ABOUT THE AUTHORS

Jean M. Larson is responsible for the development and coordination of the University of Minnesota Landscape Arboretum's Center for Therapeutic Horticulture. In addition, she teaches courses in the field of therapeutic horticulture at the University; facilitates therapeutic horticulture programs with a variety of population groups (e.g., patients with Parkinson's disease, students with chemical health issues, and intergenerational programs for elders and children together); and shares the principles of therapeutic horticulture with the general public through lectures, consulting, and publications.

Mary Hockenberry Meyer is Professor and Extension Horticulturist in the Minnesota Landscape Arboretum at the College of Agricultural, Food & Environmental Sciences (University of Minnesota). She is responsible for the development and coordination of statewide multimedia educational programs in environmental horticulture. Dr. Meyer is the State Coordinator for the Master Gardener program, which is based at the Minnesota Landscape Arboretum.

CONTENTS

Foreword   **ix**
Carla E. S. Tabourne

Chapter 1. Introduction   **1**

Objective of Sourcebook and Benefits
  of Intergenerational Gardening   1
Benefits for the Elderly   3
Benefits for the Children   4
How to Use This Sourcebook   4

**Chapter 2. Strategies for Working with Elders
and Children**   **9**

Developing Program Content   9
Promoting Interaction   11

Chapter 3. Designing Accessible Gardens   **13**

Accessible Design Standards   14
Accessible Containers   15
Constructing a Raised Bed Planter   17
Suggestions for a Sensory Garden: Choosing Plants
  That Stimulate All Five Senses   18
Basic Sensory Garden Design   21

**Chapter 4. Six-Week Intergenerational Sensory Garden
Activity Plans**   **27**

Awakening the Senses in the Garden:
  Overall Program Overview   27
Preprogram Orientation for Elders   28
Preprogram Orientation for Children   30
Week #1: Getting to Know You   31
Week #2: The Brain and the Senses   37
Week #3: Sound and the Ear   45
Week #4: Sight and the Eye   55
Week #5: The Sense of Touch   61

Week #6: The Senses of Taste and Smell 68
Postprogram Party 76
Certificate of Completion Example 76

Chapter 5. Evaluation Strategies 77

Rationale to Evaluation 77
Evaluation Example 77
Postevaluation 78

Resources 81

Basic Horticulture 81
Botanic Gardens and Arboreta 81
Honeybees 82
Horticultural Therapy 83
Intergenerational Programs 83
Sensory Gardens 84
Therapeutic Garden Program Curriculum 85
Tools 85
Universal Design 86

Index 87

Foreword

Every so often a book is discovered that becomes a cherished, much-used resource. *Generations Gardening Together: Sourcebook for Intergenerational Therapeutic Horticulture* may be the first practical resource written for recreation therapists, horticulture therapists, and Master Gardeners that is grounded in theories—specifically developmental, continuity, and exchange frameworks—and based on twelve years of curriculum development with multi-level learning outcomes. It is destined to become one of those well-worn reference books with dog-eared pages. Professionals of three disciplines—gerontology, youth services, and health care—will be able to follow explicit directions to help their clients tap into the life-enhancing properties of plants. The strategies are described in detail, but they are also streamlined to show how elders and children, with and without functional limitations, can benefit from gardening together.

Jean M. Larson and Mary Hockenberry Meyer brilliantly tackle ageism and disabilityism, two aspects of racism, and a social problem with financial, political, legal, sociological, and psychological implications for populations around the world. Addressing this problem through leisure via nonthreatening, enjoyable, and stimulating means is a major contribution to the literature. This book is written expressly for recreation therapists, horticultural therapists, and Master Gardeners; sociologists might want to examine the success of spiritual renewal with the earth, sensual awakening, remotivation, and generativity among elders, and development of industry and reciprocity with adult role models by children in the program from which the curriculum emerged. This sourcebook is a user-friendly guide for anyone wanting to experience and understand the relationship between plants and humans in sensory gardening. The curriculum has multiple learning objectives relative

doi:10.1300/5531_a

to leisure, skill acquisition, inter- and intrapersonal psychosocial learning, and spiritual and emotional development.

Intentional program outcomes for the educational curriculum are in five domains: physical, psychological, emotional, social, and spiritual. Exclusive focus on one domain is entirely possible by applying the objective steps to plan a complete series of single-theme activities based on the six-week model outlined in the book. The evaluation strategies will guide revision and redesign as the program is developed. Information presented in the included diagrams, along with the list of resources for further study, is invaluable content.

The curriculum has therapeutic outcomes for the same five domains. Therapeutic activities have been used to care for people with illness, as recorded in text and in pictures, since the beginning of civilization. Archaeologists have found evidence that gardening was one of the favored activities employed as treatment for a variety of conditions. Gardens, gardening with music and the sound of water, and flower extract aromas were described as treatments for mental illness in Greece during the 300s BC, then employed by the Romans for war distress (post-traumatic stress disorder). Larson and Meyer have captured in one book the essential components necessary for bringing elders and children of differing abilities and limitations together in a sensory garden in which many things can grow—not the least of which are knowledge, receptivity, sensitivity, and attentiveness to self and others in the dance of life.

Share this book and grow lots of ideas for activities that bring generations of mixed ability together with Mother Earth.

Carla E. S. Tabourne, PhD, TRS
Associate Professor, Recreation Therapy
University of Minnesota

Chapter 1

Introduction

OBJECTIVE OF SOURCEBOOK AND BENEFITS OF INTERGENERATIONAL GARDENING

At the heart of any intergenerational connection is the belief that each of us, at every age, has value. Intergenerational alliances engage all citizens in the healthy development of children and adolescents. Such alliances shift the paradigm of *older person* from powerless, passive, and dependent to active, creative, and productive. Cross-age connections emphasize ability; and thus, by working together, old and young can be catalysts for social change (Newman, 1989). The late Maggie Kuhn, founder of the Gray Panthers, said, "Older people have a particular responsibility in our society to develop new roles. Elders are responsible for the tribe's survival and for those who come after. We, the elders, should be society's futurists" (Kuhn, 1991).

Where does the process of change begin? For those who garden and have experienced the restorative power of plants, the garden seems a natural place to start. Horticulture has been regarded as the number one leisure activity (Simson and Haller, 1997) and provides wonderful opportunities for exercise and socialization (Haas et al., 1998; Hazen, 1997). The garden neither judges nor discriminates. It's a safe environment where people of all ages, backgrounds, and abilities can come together (McGuire, 1997; Ventura-Merkel et al., 1989; Tice, 1985; Abbott et al., 1997), connected by the simple fact that all people rely of the earth to survive (Lalli et al., 1998).

doi:10.1300/5531_01

Intergenerational gardening programs are designed to bring the generations together. These programs can benefit the elder and child, together and individually (Ward et al., 1996; Pentz and Straus, 1998; Kerrigan and Stevenson, 1997). Epstein and Greenberger (1990) report that intergenerational horticultural therapy programs are beneficial toward the development of relationships between nursing home residents and schoolchildren. In addition, horticulture therapy programs help counteract negative stereotypes and create positive images about the elderly (Ward et al., 1996). However, the empirical research into intergenerational horticultural therapy programs has not been widely studied; therefore, further research is necessary to replicate results for any conclusive outcomes (Predny and Relf, 2000).

Empirical evidence supports the general benefits of intergenerational programs. Sally Newman, of the University of Pittsburgh, has conducted a significant amount of research and organized a coalition of forces as Generations Together (see Resources section for Web site). According to Newman (1989), since the mid-1980s model intergenerational programs have been organizing at the state level and regionally. These networks include collaborations between local government and education agencies, along with health care systems, representing the elderly and children (Newman, 1989). However, the success of the intergenerational program is dependent on the planning and design process. That is, one needs to be mindful to protect the integrity of both elders and children in the group; limit the frustration for both elders and children; plan for interaction that is functional and non-contrived; and assure the interaction between elders and children is rewarding and pleasant for both (Seefeldt, 1987).

According to Carol Seefeldt (1989), a professor of education at the University of Maryland, benefits to elders in intergenerational programs, as noted in the results of several surveys, show the elderly experienced increased feelings of well-being and life satisfaction as a result of their involvement with children. The results for children, however, are not as consistent. Seefeldt's (1989) survey results were positive with regard to children's attitudes toward the elder participants. Kocarnik and Ponzetti (1991) report that both positive and negative attitudinal outcomes are identified in the research. However, the inconsistency is due to discrepancies in participant abilities, research methodologies, and types of programs being offered (Kocarnik and Ponzetti, 1991; Predny and Relf, 2000).

BENEFITS FOR THE ELDERLY

Anecdotally, the benefits of intergenerational gardening with the elderly seem apparent.

- For example, the elder benefits because the garden is a place where one can promote positive self-esteem. That is, through the garden a person can view himself or herself as someone with a history of strength, competency, and productivity. This is particularly important as a person ages and feels more dependent on others.
- Gardening can inspire elders to continue to use their brains. Gardening makes us think as well as feel. When elders cannot remember recent events or situations, they may feel incompetent. Gardening can restore a feeling of competence because it uses both long-term and short-term memory skills.
- Gardening can instill pride in one's abilities. Anyone can admire a planter filled with beautiful flowers. But a sense of satisfaction emerges when one sows a seed and ends up with a beautiful and productive plant.
- Gardening brings comfort to a person, especially if he or she attaches particularly fond feelings to certain plants and/or plant scents. They may trigger memories of people, places, and/or vocations.
- Through gardening, an elder person can pass on his or her knowledge to the younger generation. Through gardening, the elder provides a legacy of wisdom for future generations (and/or a legacy of future plants with heirloom seeds and newly planted trees).
- Gardening is physically stimulating. It is common for many people to feel more physically alert and healthy after gardening. Gardening is a lifelong leisure skill to carry throughout life.

BENEFITS FOR THE CHILDREN

Likewise, the intergenerational gardening experience can benefit children in the following ways:

- The garden is a safe environment where people can come together to learn about plants and how they benefit humans.
- The garden is the agent by which youth can seek the wisdom of elders. Not many opportunities are available for elders and children to interact—but a garden is an ideal vehicle for social interaction.
- The garden is the place where youth can learn lessons of accountability, nurturing, and responsibility. The garden teaches about life, death, hope, patience, and beauty. It connects youth to the land. It provides young people a place to explore, rejoice, and learn about their connection to living things.
- Last, the garden's recurring cycles reassure us all of life's continuity.

In general, the prevailing principle of intergenerational gardening for therapy is to benefit and improve one's physical, emotional, and spiritual well-being. It encourages participants, young and old, to use their senses while working with plants. It allows people to influence their environment in a positive manner. It encourages creativity. It provides for both individual and group accomplishments and self-assurance.

It is the intent of the authors of this sourcebook to encourage others to use the garden as a catalyst for change within their communities, states, nations, and even around the world. When intergenerational programs are perceived as strengthening communities, they garner public support and demonstrate the interdependence of all generations.

HOW TO USE THIS SOURCEBOOK

This sourcebook has been created for health care professionals, horticultural therapists, and Master Gardeners. These horticultural-therapy-based activities will assist anyone interested in a successful intergenerational gardening experience.

Over a period of ten years of summer intergenerational programs at the Minnesota Landscape Arboretum, elders and children have created, tested, and evaluated the activities complied in this sourcebook. The activities rely on inexpensive, readily available tools and resources that are easily obtainable throughout the growing season.

Chapter 2 of this sourcebook begins with an introduction to the strategies of working together with elders and children. Chapter 3 then introduces the concept of accessible garden design, featuring a sensory garden for conducting the intergenerational garden program.

The heart of the sourcebook, Chapter 4, consists of a six-week program curriculum that can be used to facilitate an intergenerational program. A preactivity session is recommended, for a total program length of seven weeks. The activities are recommended for use with a group of eight to ten participants for one hour each week. Ideally, an equal ratio of elders to children (with four to five of each age group) is best. The youth should be between ages eleven and thirteen. Elders could include anyone over age seventy. It is recommend as a six-week program, but it can be adapted to fit individual group needs. We encourage you to view the sourcebook as a flexible structure to fit your own creative imagination. Read through each section and find the components that best suit your needs and adapt and modify them accordingly.

The sourcebook concludes with Chapter 5, an evaluation section to determine the accomplishments and/or improvements necessary to add to or subtract from your program. Also, a useful Resources section, based on the personal experience of the Minnesota Landscape Arboretum horticultural therapist, follows the text. Thank you to Sara Thoms for her articulate and precise work on the index and final proofreading of the text. We could not have done it without her assistance.

Prerequisites

1. It is assumed that the user of the sourcebook has some working knowledge of how the disabling conditions of elders can affect their function, as well as how to assess participant skill level to formulate therapeutic goals and objectives.

2. In addition, it is assumed that a health care provider will be able to understand the safety issues of the participants and how to effectively adapt and modify the program and equipment to maximize the participants' independence.

3. Last, it is necessary to have an accessible garden space (that is, a garden near or adjacent to paving which will allow users who depend on wheelchairs, walkers, canes, or electric scooters) that receives at least eight hours of sunlight each day and is near a water source, shelter, and toilet facilities.

Format of Activity Sessions

Each activity session includes the following information:

Week and Title: Represents the content of each weekly program
General Statement of Purpose: Identifies the intent of the program
Goal(s): The expected outcome(s) of the activity being delivered
Materials and Equipment: Identifies all the necessary equipment and supplies needed to facilitate the program activity
Program Procedures: Includes a detailed description of steps and order of program activity
Evaluation: Includes what and how therapeutic program goals are to be measured and recorded
Diagram(s) and/or Handouts: Full-page exercises to be photocopied and distributed

REFERENCES

Abbott, G., Cochran, V., and Clair, A.A. (1997). Innovations in intergenerational programs for persons who are elderly: The role of horticultural therapy in a multidisciplinary approach. *Activities, Adaptation and Aging*, 22(1/2), 27-37.

Epstein, S.G. and Greenberger, D.S. (1990). Nurturing plants, children, and older individuals: Intergenerational horticultural therapy. *Journal of Therapeutic Horticulture*, 5, 16-19.

Haas, K., Simson, S.P., and Stevenson, N.C. (1998). Older persons and horticultural therapy practice. In S.P. Simson and M.C. Straus (Eds.),

Horticulture as therapy: Principles and practice (pp. 231-255). Binghamton, NY: The Haworth Press.

Hazen, T.M. (1997). Horticultural therapy in the skilled nursing facility. *Activities, Adaptation and Aging,* 22(1/2), 39-60.

Kerrigan, J. and Stevenson, N.S. (1997). Behavioral study of youth and elders in an intergenerational horticultural program. *Activities, Adaptation and Aging,* 22(3), 141-153.

Kocarnik, R.A. and Ponzetti, J.J. (1991). The advantages and challenges of intergenerational programs in long term care facilities. *Journal of Gerontological Social Work,* 16(1/2), 97-107.

Kuhn, M., Long, C., and Quinn, L. (1991). *No stone unturned: The life and times of Maggie Kuhn.* New York: Ballantine Books.

Lalli, V.A., Tennessen, D. J., and Lockhart, K. (1998). *Using plants to bridge the generations: Horticulture and intergenerational learning as therapy.* Ithaca, NY: A Cornell Cooperative Extension Publication.

McGuire, D.L. (1997). Implementing horticultural therapy into a geriatric long-term care facility. *Activities, Adaptation and Aging,* 22(1/2), 61-80.

Newman, S. (1989). A history of intergenerational programs. *Journal of Children in Contemporary Society,* 20, 1-15.

Pentz, T. and Straus, M.C. (1998). Children and youth and horticultural therapy practice. In S.P. Simson and M.C. Straus (Eds.), *Horticulture as therapy: Principals and practice* (pp. 199-230). Binghamton, NY: The Haworth Press.

Predny, M.L. and Relf, D. (2000). Interactions between elderly adults and preschool children in a horticultural therapy research program. *HortTechnology,* 10(1), 64-70.

Seefeldt, C. (1987). Intergenerational programs: Making them work. *Childhood Education,* 64(1), 14-18.

Seefeldt, C. (1989). Intergenerational programs: Impact on attitudes, *Journal of Children in Contemporary Society,* 20, 185-194.

Simson, S. and Haller, R. (1997). Horticulture therapy education and older adults. *Activities, Adaptation and Aging,* 22(3), 125-140.

Tice, C.H. (1985). Perspectives on intergenerational initiatives: Past, present, and future. *Children Today,* 14(6), 7-11.

Ventura-Merkel, C., Liederman, D.S., and Ossofsky, J. (1989). Exemplary intergenerational programs. *Journal of Children in Contemporary Society,* 20, 173-180.

Ward, C., Kamp, L., and Newman, S. (1996). The effects of participation in an intergenerational program on the behavior of residents with dementia. *Activities, Adaptation and Aging,* 20(4), 61-76.

Chapter 2

Strategies for Working with Elders and Children

Before you get started, here are some keys to success strategies to consider when working with elders and children together.

DEVELOPING PROGRAM CONTENT

Intergenerational garden programs are successful when careful consideration is given to the planning and development of program content. Following are a few guidelines for working with elders and youth together.

1. *Orientation*—Establish separate orientations for the elders and for the children before the program starts. This will help clarify expectations and dispel any myths about working together (see Preprogram Orientation for Elders, Chapter 4).
2. *Avoid stereotypes*—Expose children to a variety of elders in the group to reduce unintentional reinforcement of certain age-related stereotypes, such as the notion that all older people are frail and disabled. Such exposure to diverse members of an age-related group will deter the tendency to generalize a small sample of the individual characteristics to an entire group of people.
3. *Name tags*—Wear name tags at all times. This may seem simple, but it will help promote interaction when people know one another's name.

doi:10.1300/5531_02

4. *Emphasize ability*—The garden acts as catalyst of the "can-do" attitude. Bringing young and old together emphasizes and complements both groups' needs, for example, where one lacks strength, the other has it.

5. *Themes*—Create garden themes that reflect participant interest—for example, ethnic foods, bird and butterfly gardens, or planting for wildlife.

6. *Grandparent buddies*—Establish "buddies" at the onset of the program that will stay the same throughout the duration. Buddies allow for consistent friendship building and opportunity to grow together and interact on a regular basis.

7. *Developmentally appropriate activities*—Conduct activities which the whole group is interested in doing and which are developmentally appropriate. Take into consideration participants' strengths as well as their functional limitations. Ask them, "What do you want the horticulture project to look like in one month (one season, one year)?" Brainstorm activity ideas. This can be accomplished with a simple interest assessment.

8. *Success*—Plan reasonable goals and objectives with reasonable timelines for completion and success. It is important to have an overall long-term plan to manage the short-term weekly operations.

9. *Stimulate interaction*—Structure for interaction through use of cooperative goals, in other words, assign tasks so that everyone in the group plays a role in the project's success (see the next section on promoting interaction). One predominant theme that runs throughout the activities is interaction. It is important that you, as the facilitator, promote question asking, encourage discussion about similarities/differences in participants' answers, and provide ongoing encouragement for participants to share together what they learned from the activities.

10. *Promote learning*—Take time to reinforce the learning outcomes and life skills developed during the activities. Make sure all staff, volunteers, and participants are clear about the academic goals with each activity. Include a time for reflective

processing at the end of each session activity to discern what was learned during the time spent in activities.

11. *Have fun and celebrate success!*—The program will be a success when the facilitator, elders, and youth are all engaged in the activity and the project culminates with a celebration of completion.

PROMOTING INTERACTION

Following are a few examples of additional programs and activities with cooperative goals that produce positive results in social interaction.

Scavenger Hunt

Ask participants to work as partners with blindfolds on a sensory hike.

Poetry

Have group members use words that describe an object or action in nature; ask them to try to make haikus by counting five syllables for the first line, seven for the second, and five for the last line, for example,

> Windy breeze cools me,
> Shade trees big, small, green, and tall,
> Many friends share all

Progressive Lunch

Each pair of participants make a different item for a progressive lunch event (e.g., herb butter with crackers, herbed cream cheese with bagels, herbs and goat cheese pizza, edible flower salad, herbal iced tea, or edible flower cookies and bars).

Show and Tell

Each group finds a plant and looks up information about it in a variety of plant resources (such as encyclopedias or gardening handbooks) and then shares what has been learned with the others in the group.

Flower Power

Each group member adds a flower to a flower arrangement to be given to someone special.

Wonders of Worms

Each group has a role in constructing a worm bin. Activity culminates with "compost pudding"—an edible treat of crushed Oreo cookies mixed with chocolate pudding and gummy worms.

Garden Gazette

Each child asks his or her elder partner a list of questions about gardening. All information is gathered together for an end-of-the season-newsletter. Some questions could include the following:

> What is your earliest memory of gardening?
> What were your favorite foods?
> How did you shop for food?
> How did you store food?

Creature Feature

At the end of the season, hold a harvest festival. One of the events could be for the partners to create "creatures" using the over-sized zucchini and leftover produce.

Chapter 3

Designing Accessible Gardens

Before you read this chapter, let us say that many resources are available which pertain directly to accessible garden design. Please refer to the Resources section at the end of this book for a list of the books and Web sites that go in depth on the subject. However, in general, an accessible garden design makes good sense for anyone who gardens because of the therapeutic benefits. The key to an accessible design is good site planning, that is, blending individual needs and abilities with environmental considerations. Therefore, the site will complement, rather than compete with, the location and the people it serves.

Here are some things to consider when designing accessible garden spaces:

- *Allow for multiple levels of individual ability and/or access.* If not everyone uses a wheelchair, then provide space for wheelchair access but create a variety of heights, levels, and opportunities for everyone to enjoy.
- *Provide access to all areas of the garden.* Make pathways wide enough for people using wheelchairs and stable enough for people who use walkers and canes. Make the garden paths shorter and arrange the area in smaller blocks to make the garden easier to reach.
- *Provide proximity to garden facilities.* Arrange water, storage shed, fountain, compost pile, etc. in a central location where all can be readily accessed from the four corners of the garden.
- *Balance accessibility with challenges.* You must make the garden safe, and you must consider the ability levels of the patrons

of the garden area; however, remember that people come to the garden to gain self-confidence and independence, and they can be responsible for their own actions. Therefore, when designing a garden space, create a space where one can work within individual limits. A well-deigned garden is made accessible and effective for the person with limitations but does not impinge on the individual's choice to challenge those limitations.

ACCESSIBLE DESIGN STANDARDS

The following list contains some physical factors that should be taken into consideration when designing an accessible garden.

1. Water must be available close to the garden site and in a paved area so the ground does not get muddy. Place the spigot twenty-four to thirty-six inches aboveground and use hand levers, not round spigot handles, and snap connectors. Soaker hoses and mulch will also reduce watering needs in the garden.
2. To aid in harvesting plants, use contrasting or bright colors. Some plants naturally contrast their ripening fruit against their foliage, such as purple-podded bush beans or golden zucchini. Select plants with interesting textures and fragrances and those that are high producers per inch of growing space. Use plants that people want to grow or eat.
3. Provisions should be made to summon assistance for potential medical or police emergencies. A wheelchair-accessible parking space near the garden is mandatory for public gardens, both for persons with disabilities and for medical/police access. It is not mandated at private homes.
4. Garden path surfaces must be firm, smooth, level, and provide traction. The grade of the path should be no greater than 5 to 8 percent. Provide direct routes throughout the garden, with textured edge guides if you have participants with ambulation and/or visual disabilities. Audible water features and wind chimes also help guide participants through the garden. One-way traffic needs a minimal width of five feet to

accommodate the turning radius of a wheelchair. Two-way traffic requires a minimal width of seven feet.

ACCESSIBLE CONTAINERS

Accessible containers come in all sizes, shapes, materials, and heights. A few to consider when designing an accessible garden are listed here.

Raised Beds

Raised beds are large bottomless boxes that contain soil and permit drainage below. They can be expensive to build, so raised beds should be used in areas of the garden that require the most frequent attention. For instance, it makes more sense to build a raised bed for vegetables requiring intensive weeding than for a low maintenance border of shrubs. Build raised beds as large as possible, making sure that your participants can reach all areas of the bed. The increase in size adds minimal cost to the bed, while adding valuable garden area. Bed width should be a maximum of five feet if it is accessible from all sides or two and one-half feet if used from only one side. If using extended tools, you can add inches to the bed. Seating ledges should be from eight to eighteen inches wide. Use the thinnest construction materials possible without compromising stability, to increase the area available for the gardener. Height of the sides can vary from eighteen inches for a child, to twenty-four inches for someone seated in a chair next to the bed, to thirty inches or higher for the standing gardener who has difficulty bending down low.

Boxes and Pots

Various sizes of boxes and pots can be a very successful way of growing vegetables and flowers. As a basic guideline, choose a pot that will allow for healthy root development. For instance, bush type peas, beans, cucumbers, kale, broccoli, and lettuce do well in a box that is one foot by four feet and eight inches deep. For some other plants, such as beets, carrots, onions, lettuce, leeks, turnips, kohlrabi, corn, and zucchini, a box that is two feet by three feet and ten inches deep is more suitable. For herbs and flowering plants and vines, find out whether the plant is deep or shallow rooted to determine the proper container size. The more shallow the container, the faster it will dry out.

Hanging Baskets

Use hanging baskets to create planting space where none exists. Combined with a container garden, they can provide a double-decker growing area. To make watering and viewing easy, buy a ratchet pulley or make your own. Steel hooks or rings can be clamped or mounted to railings or walls. A long metal pole with a curving top hook can be anchored in the ground for a freestanding hanging plant mount. Baskets can be hung high or, for those with limited mobility, hung low enough to see and enjoy their beauty.

Table Planters

Table planters are shallow soil-filled trays supported on legs. About twenty-seven inches of knee clearance is needed to allow chairs to fit underneath. The soil container should be at least eight to ten inches deep, making the entire structure about thirty-five to thirty-seven inches high. If standing, the top of the planter should be no higher than an adult's rib cage. The width of the box is the same as that of the raised bed, described previously.

Deep Boxes, Barrels, and Tubs

Deeper containers can be used to create miniature raised beds for annual flowers, vegetables, and herbs. Perennials, trees, and shrubs are not recommended for these types of containers due to the possibility of freezing and the negative effects of extremely cold temperatures on plant roots. Use these containers for annual plants only.

CONSTRUCTING A RAISED BED PLANTER

Raised beds make gardening accessible for everyone and are better for back health. The following instructions explain how to build a basic wooden raised bed. Customize the raised bed for individual or group needs, abilities, and desires.

Site Selection

The raised garden bed is aboveground; therefore, the condition of the soil under the bed is not important. However, the surface surrounding the bed should be level and accessible for those who will use it. Consider, for example, grass: Although grass is attractive, it may be difficult for someone who uses a wheelchair to access the gardening area around the raised bed. Consider placing a raised bed on or near concrete, which can enhance a patio or sidewalk. It will also stay weed free and easily accessible to all.

Construction

These instructions are for an open box, approximately two feet high by four feet wide (by any length desired). In choosing the length, consider that lumber comes in standard lengths of eight, ten, and twelve feet. A maximum width of four feet is recommended to enable the gardeners to easily reach the middle of the bed.

1. Using two-inch-by-eight-inch boards, lay out the bottom layer of boards in a rectangular box-shaped frame.
2. Attach the corners with screws (use two-and-one-half-inch drywall screws or three-inch deck screws) at each corner.
3. Make two more box-shaped frames the same way.
4. Stack the three frames to make a twenty-two-and-one-half-inch-tall bed.
5. Secure two-by-four braces in each corner with screws to tie the three frames together. Both sides of the frame should be securely attached to the corner braces to withstand the

outward pressure of wet soil over time. (Add an extra row of screws at each corner to secure the end pieces to the braces.)
6. For sides longer than four feet, attach a two-by-four to the inside with screws for every three feet in length.
7. If the bed is not on a hard surface, make the two-by-four longer and stake it in the ground before attaching, to keep the sides from bowing out.

Soil

Once the bed has been constructed, the bottom third of the box can be filled with crushed rock or gravel to allow for good drainage and reduce the amount of soil needed. Next, fill the bed with soil or other planting media, adding extra to allow for settling. Use a planter mix, or make your own mix of one part garden soil to one part sand to one part compost.

Materials

Use naturally rot-resistant wood such as redwood, cedar, or plastic timbers. For food production, treated wood of any kind can be dangerous and is not recommended. To help extend the life of the wooden raised bed, line the sides with thick black plastic. Besides wood, many options are available for raising the soil level, including several raised-bed kits found in garden catalogs, home improvement stores, and garden centers (see Resources section).

SUGGESTIONS FOR A SENSORY GARDEN: CHOOSING PLANTS THAT STIMULATE ALL FIVE SENSES

Plants That Provide Fragrance or Smell

Annuals	Scientific Name
Chocolate cosmos	*Cosmos atrosanguineus*
Heliotrope	*Heliotropium arborescens*
Herbs	Any kind
Marigold	*Tagetes* spp.

Pansy   *Viola wittrockaina*
Sweet alyssum   *Lobularia maritima*

Vines
Moonflower   *Ipomoea alba*
Sweet pea   *Lathyrus odoratus*
Trumpet honeysuckle   *Lonicera sempervirens*

Perennials
Lily of the valley   *Convallaria majalis*
Peony   *Paeonia lactiflora*

Bulbs
Hyacinth   *Hyacinthus orientalis*
Lily   *Lilium* spp.

Shrubs
Lilac   *Syringa vulgaris*
Rose   *Rosa* spp.

Trees
Apple   *Malus* spp.
Balsam fir   *Abies balsamea*
Littleleaf linden   *Tilia cordata*

Plants That Are Interesting to Feel or Touch

Annuals   Scientific Name
Dusty miller   *Senecio cineraria*
Fountain grass   *Pennisetum* spp.
Rose-scented geranium   *Pelargonium graveolens*

Perennials
Lamb's ears   *Stachys byzantina*
Silver mound artemisia   *Artemisia schmidtiana* 'Silver Mound'

Trees

Amur chokecherry	*Prunus maackii*
River birch	*Betula nigra*
Bur oak	*Quercus macrocarpa*
White pine	*Pinus strobus*

Plants with Auditory Features or Sound

Wind through trees and grasses—the susurration or whispering of evergreens, quaking aspen, and tall grasses (such as *Panicum* and *Miscanthus*)—these plants all make lovely sounds when the wind rustles through their foliage.

Other Natural Sounds to Enjoy

Just as plants can be selected to provide sensory stimulation (such as sound) so can nonplant garden elements. Choosing plants that bring in helpful pollinators, such as bumble bees, honeybees, or mason bees is one way to provide additional sensory stimulation. Also, create habitats (that include food, water, shelter, and space) to encourage birds to stay in the area. Incorporating a water feature or planting a mini-wetland habitat can bring in a wide variety of birds and insects. Or, simply stand outside in a protected area during a rainstorm to hear the thunder roar. Don't forget to attach a wind chime to catch the wind and remind you of the beauty found in the invisible.

Chirping crickets
Buzzing bees (bumble, honey, and mason are excellent additions to a sensory garden)
Singing birds
Rain and thunder
Fountain or waterfall
Wind chimes

Edible Plants

The best part of gardening is eating—which can be done with fruits, vegetables, and herbs. The world is full of varieties, so decide on a kind, then plant it, take good care of it, and enjoy the fruits of your labor! Engage the group in a discussion of what plants should be grown or purchased to eat as a class. Fast-growing vegetables such as lettuce, Swiss chard, and spinach are easy, or purchase herbs already large enough to enjoy with other vegetables. Rosemary with potatoes, basil with tomatoes, mint and ice tea are favorite combinations.

Remember to avoid toxic plants and be aware of those with stickers and thorns.

BASIC SENSORY GARDEN DESIGN

The following figures and photos are based on the Clotilde Irvine Sensory Garden at the Minnesota Landscape Arboretum.

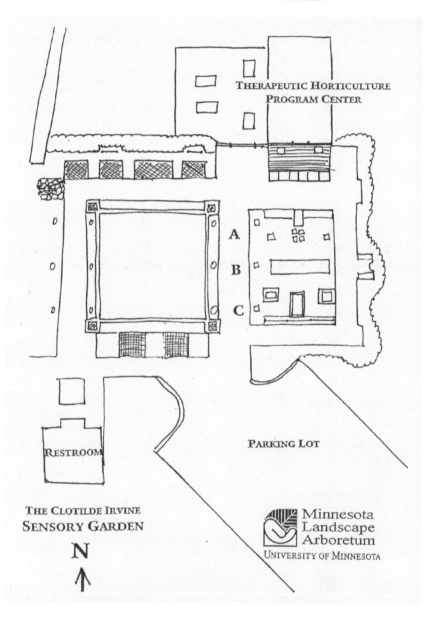

FIGURE 3.1. Welcome to the Minnesota Landscape Arboretum's Clotilde Irvine Sensory Garden and Therapeutic Horticulture Program Center. This garden was designed to tickle the senses and keep your interest all year long.

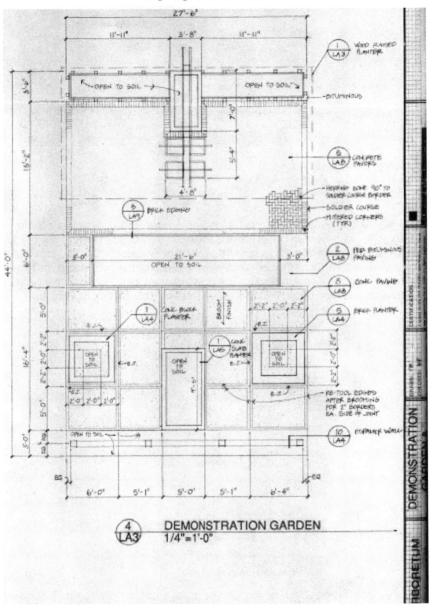

FIGURE 3.2. The Demonstration Garden, a close-up of garden sections A, B, C, includes containers, paving materials, and design features for full accessibility. Further detail of these areas is given in the following illustrations.

The Idea Garden of the Minnesota Landscape Arboretum's Clotilde Irvine Sensory Garden. The Idea Garden of the Clotilde Irvine Sensory Garden is a display of containers, paving material, and design features and is fully accessible. The area offers a variety of ideas which will make gardening more accessible and enable you to garden beyond conventional methods. The plants chosen for these beds change year to year and are used by participants in the Therapeutic Horticulture Program.

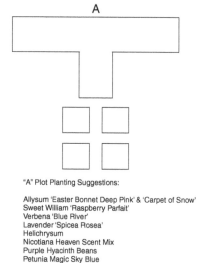

A

"A" Plot Planting Suggestions:

Allysum 'Easter Bonnet Deep Pink' & 'Carpet of Snow'
Sweet William 'Raspberry Parfait'
Verbena 'Blue River'
Lavender 'Spicea Rosea'
Helichrysum
Nicotiana Heaven Scent Mix
Purple Hyacinth Beans
Petunia Magic Sky Blue

FIGURE 3.3. This section of the demonstration garden is a large, T-shaped, raised bed with a tall, permanent wooden trellis placed at right angles within the bed and four large chimney flues; see right secton of photo. It is planted with flowers noted for fragrance and color. The specific cultivars listed are used to complement one another and create a soft and calming effect.

B

"B" Plot Planting Suggestions:

Nicotiana 'Fragrant Cloud'
Allysum Carpet of Snow
Snapdragon Royal Bride
Snapdragon La Bella Pink
Lavender Munstead
Petunia Magic Sky Blue
Dianthus Plumarius 'Sonata'
Penisetum Rubrum

FIGURE 3.4. This is a standard ground bed with a brick edging for demarcation. It also has a variety of plants chosen for texture and fragrance.

Springtime at the Minnesota Landscape Arboretum's Clotilde Irvine Sensory Garden.

C

Nicotiana 'Heaven Scent'
Allysum Easter Bonnet Deep Pink
Sweet William 'Raspberry Parfait'
Rosemary Lavandulaceus (Prostrate)

Snapdragon 'Royal Bride'
Helichrysum
Petunia Magic Sky Blue
Salvia 'Sage of Bath' 18"

Verbena 'Blue River'
Allysum 'Carpet of Snow'
Heliotrope 'Marine'
Lavender Munstead

FIGURE 3.5. Each of these containers is made from concrete-type materials such as pavers, blocks, or stacked bricks. All have a ledge for easy, seated access; see the left section of the photo.

Chapter 4

Six-Week Intergenerational
Sensory Garden Activity Plans

AWAKENING THE SENSES IN THE GARDEN:
OVERALL PROGRAM OVERVIEW

Audience

Elders (seventy to eighty years old)
Children (eleven to thirteen years old)

Overall Program Purpose

1. To learn about human senses and their function
2. To promote social interaction between elders and children
3. To teach basic horticultural science

Program Schedule

Preprogram: Orientation for elders; orientation for children
Week #1: Getting to Know You
Week #2: Introduction—The Brain and the Senses
Week #3: Sense—Sound and the Ear
Week #4: Sense—Sight and the Eye
Week #5: Sense—The Sense of Touch
Week #6: Sense—The Senses of Taste and Smell
Postprogram—Party

doi:10.1300/5531_04

Location

This program has taken place over the past ten years at the Minnesota Landscape Arboretum Clotilde Irvine Sensory Garden and Program Center. It can be facilitated anywhere that children, elders, and gardens can be brought together.

Participants

Over the past ten years, the elders have participated from the Sojourn Adult Day Care Center located in Spring Park, Minnesota. Participants have been diagnosed with a variety of ailments, including Parkinson's disease, Alzheimer's disease, strokes, and general aging-related limitations (poor eyesight, difficulty with hearing, limited range of motion, decreased stamina, etc.)

The children who participated came from the Excelsior summer community recreation program called The Wa-Cha-Ma-Call-Its. They come from a variety of abilities, cultural and socioeconomic backgrounds.

PREPROGRAM ORIENTATION FOR ELDERS

The elders were given a brief thirty-minute orientation to the summer's program at the Sojourn Adult Day Care. At the beginning of the session, the elders were asked to share, in single-word phrases, how they would describe children in today's world. Some of the words expressed included the following:

Crazy
Mean
Drugs
Weird
Rude
Disrespectful
Lazy
Spoiled

Then the elders were asked to recall words that their parents and others used to describe them when they were children. This time they used such words as these:

Obedient
Quiet
Loony
Disciplined

Then, the elders were asked to describe the children in their own lives (such as their grandchildren, great-grandchildren, nieces, and nephews). These children were described as follows:

Sweet
Caring
Generous
Special

All three of these lists were generated on large pieces of paper and written large enough for the elders to read clearly. The papers were then taped to the wall. Discussion was generated which suggested we tend to objectify and sterotype individuals or groups that we do not know or have personal relationships with. The opposite is true when we know the person or have a personal relationship: the words become more subjective and tolerant. It was made clear through stories and experience that by the end of the summer the words the elders would use to describe the children in the summer program would be more similar to those they chose to describe their own family. After a full six weeks, some "surrogate" grandchild relationships would have developed.

Afterward, the elders watch a video from the past summer's horticultural therapy program that highlighted the interactions between the elders and the children. The elders recognized the children and one another in the video.

The program ended with hopes for new relationships and good cheer.

PREPROGRAM ORIENTATION FOR CHILDREN

The children were given a brief thirty-minute orientation at the Excelsior Community Recreation Center. During the thirty minutes, the children watched a video from the past year's horticulture therapy program that highlighted the interaction between the elders and the children.

Afterward, the children were asked to draw pictures of what they perceived an old person to look like. These pictures typically included gray hair, glasses, outdated clothes, and no smiles on the old person.

Then, the children were asked to draw pictures of their own grandparents, or aunts and uncles if they did not have a grandparent. Here the drawings had smiles; most still had gray hair, but the elders were often smiling and the children were sometimes included in the picture.

Next the children were asked to compare and contrast the two drawings—one of "an old person" and the other of their own older loved ones. Discussion surrounded the similarities and differences.

Then each child was asked to draw a picture of himself or herself along with an older person he or she would like to be partnered with this summer. The pictures included their smiling faces along with an older smiling person. The children were assured they would find, over the course of the summer, a new friend in their elder partner.

Afterward, the children were introduced to the definitions of some of the more common ailments the elders in the summer would be dealing with (such as Alzheimer's, Parkinson's, and strokes).

The children were asked if they knew of anyone with any of these types of diseases or conditions and, if so, if they could describe what they see and understand about them.

The children were then presented with some common definitions:

Alzheimer's disease: Named after Alois Alzheimer, a German neurologist, this disease is a kind of organizing mental disorder that describes a general loss of cognitive ability. The loss can be observed as a loss in long-term memory, judgment, and abstract thinking, and as changes in personality. In other words, the person changes significantly from being able to care for himself or herself to being dependent on others to help cope with some of the most simple tasks of daily living.

Parkinson's disease: Named after James Parkinson, this disease is known to progressively affect sensory motor coordination, causing individuals to have difficulty initiating activity. It is seen typically in muscular tremors and peculiar gait.

stroke: Strokes can occur when someone has a restricted blood supply to some part of the brain. When the brain cells do not receive the oxygen contained in the blood, they start to die. The duration and location of the restricted blood flow determines the stroke outcome. For example, a person may not have use of upper body limbs; may have loss of balance or coordination; or may experience slurred speech or decreased vision.

We concluded the time together with the children trying out the elders' wheelchairs, walkers, canes, and transit carts to acquaint them with and demystify these pieces of equipment. A pep talk at the end of the discussion was intended to motivate and prepare children for their experiences with the elders the next week.

WEEK #1: GETTING TO KNOW YOU

Length of Time

Each weekly session is a one hour in length.

General Statement of Purpose

The purpose of this activity is to bridge the gap between elders and children through group participation and partnerships.

Goal

To increase social interaction between elders and children.

Materials and Equipment

- M&M's candies
- Name tags
- Colored pencils or pens
- Large pieces of paper or a chalkboard
- Masking tape (to hang paper)
- Multiple copies of "Getting to Know You" questions (end of this section)
- Drawing paper or scratch paper for each participant
- Notebook journals for each participant
- Pencils

Program Procedures

Name Game

Ask each member each of these questions:

- What is your name?
- How did you get your name?
- Who are you named after?
- Do you have any nicknames?
- Have you ever wished to be named something else? If so, what?
- Can you think of something you like that begins with the first letter of your first name (e.g., "My name is Jeannie, and I like jelly beans")?

Ask each member of the group to answer these questions, one after another. Repeat until everyone has said his or her name and has answered all questions.

M&M's Game

Pass out M&M's candies to each participant. Ask each person to pick out one piece of candy and describe a personal story. Each color represents a different story description.

Red: Embarrassed
Blue: Sad
Green: Jealous
Brown: Travels
Yellow: Happy

For example, I choose a red candy, so now I must tell a story of something that was embarrassing for me. Proceed throughout the group until all have done this.

"Getting to Know You" Game

Ask each group member to pick a number between one and forty. The first member announces his or her number, and the leader asks the corresponding question from the list at the end of this section. Each person in turn announces a number and answers a question.

Draw Your Name

Ask all members to draw something about themselves on their name tags. Each group member then explains his or her own drawing.

Walk in the Garden

With a partner, group members walk out into the sensory garden and find a flower that both members like. They then rejoin the larger group and tell all members of the group the name of the flower and why they chose it as the one they liked best.

Evaluation

At the end of the session, after each pair has finished explaining their favorite flower, pass out new notebook journals and pencils to each member of the group. Ask both partners to write their names on the journals. Wrap up the session by asking them to write down their thoughts to the following questions:

1. What is your partner's name?
2. What did you enjoy most about getting to know your partner?
3. What did you learn about your partner today that you will remember?

Afterward, thank them for a great program and their attention, and remind them of how important it will be in the next five weeks for the partners to work together. Tell them, "Don't forget your partner's name because you will be paired together again next time."

Week #1: "Getting to Know You" Questions

1. What is the best movie you have seen in the past year?
2. What is the most beautiful flower you have ever seen?
3. What is the ugliest flower you have ever seen?
4. What do you like to do for hobbies?
5. What traits do you most find attractive in people?
6. What is the biggest problem facing the United States today?
7. How do you feel when you are mad?
8. What is a talent that you possess?
9. What is the greatest value that guides you in life?
10. Who taught you the values that guide your life?
11. What are the qualities that you look for in a friend?
12. Who is the most influential person in your life? Why?
13. What is your security blanket?
14. How do you waste resources?
15. How can you not waste resources?
16. What is your greatest joy?
17. What is your greatest fear?
18. Select one word that describes all people your age.
19. If you could give me advice, what would you tell me?
20. What is the most unreasonable thing your parents ever did to you?
21. If you could choose to be a character in a book, who would you be and why?
22. If you could paint a picture with you as the main focus, what would you paint?
23. What do people like best about you?
24. What is your greatest struggle?
25. When do you feel most happy?
26. What bothers you on TV?
27. What is one thing you would change in your life?
28. Describe your feelings about family vacation?
29. What do you look forward to in your lifetime?
30. What do you like to talk about with your friends/family?
31. What is the very last thing you could give up?
32. What is the best advice you have ever gotten?
33. Who told you this advice?
34. What makes you feel mellow?
35. What makes you feel excited?
36. What sound would you be if you were a sound?

37. What is your favorite time of year and why?
38. What character on TV or in the movies do you identify
    with and why?
39. What are you looking forward to today?
40. What are you looking forward to tomorrow?

WEEK #2: THE BRAIN AND THE SENSES

General Statement of Purpose

The purpose of this activity is to introduce how the human brain works and the different ways that the senses are connected to the brain.

Goals

1. To introduce elementary neurological science
2. To encourage social interaction

Materials and Equipment

- Large pieces of paper or chalkboard with writing utensils
- Multiple copies of the Brain Diagram (end of section)
- Four large trays
- Neosporin antiseptic
- Bandages
- Small scissors
- Tweezers
- Cotton ball
- Pencil
- Paper
- Wite-Out (liquid correction fluid)
- Eraser
- Paper clips
- Jackknife
- Small rope
- Candle
- Insect repellent
- Matches
- Rollers
- Hair spray
- Comb
- Brush
- Barrette
- Colored pencils

Program Procedure

Introduction to the Brain and the Senses

1. Ask the participants, "What is a brain?"
2. Write down their answers on a large piece of paper or chalk-board. Make sure that their answers include the following points:

   • It is the control center for the body.
   • It stores memories.
   • It allows us to talk.
   • It controls our movements.
   • It controls our ability to solve complex problems.
   • It is in charge of our senses.

3. Remind the group that last week, when playing the "Getting to Know You" game, the capacity to play those games lay in the brain. As they remember what they did last week, explain to them that all the information stored in their memory is held in a place known as the *limbic system,* sometimes called the *hippocampus.* When a memory forms in the brain, it becomes a short-term memory, which just lasts a few minutes. The sense organs send electrical messages to the brain and cause tiny amounts of chemicals to squirt into the spaces between the nerves, allowing the messages to flow along many different paths.
4. Show and discuss the Brain Diagram (at end of this section). Discuss the diagram parts using letters; the diagram will be colored at the end of session by the participants.
5. Discuss: The brain has thousands of messages racing around inside. If a nerve cell is triggered over and over again, the results may be different. That is, the nerve cell grows and makes a new connection from the smaller short-term memories; this is called a long-term memory. So, memories depend on the connections between nerve cells and the chemical signals (neurotransmitters) to pass between them.

"So, what if a person cannot produce enough of the chemical signals that pass between the nerve cells? Memory is harmed. That is what often happens with people with Alzheimer's disease.

"Does anyone here have Alzheimer's, or do you know of anyone with Alzheimer's?

"Alzheimer's disease is when a person cannot produce enough of the neurotransmitters to make the memory connection in the brain. Can you imagine what that might feel like? Not being able to remember something that you did just a few moments before? Or the day before? That would be difficult, wouldn't it?"

Activities

Limbic System Challenge

The limbic system has two systems within it:

1. *Implicit system*—The nerves to learn actions
2. *Explicit system*—The nerves to learn people, places, and things

"Explain, these two systems work together so we can recognize and respond appropriately. So, for example, if we see a lion, our explicit system sends us signal telling the implicit system to *run!*"

Tell the group you are going to draw some situations. Then ask them to describe what the situation is and what word would describe their reaction to it. The explicit will be drawn in a picture on board; the implicit will be a word to describe the reaction to the situation.

Explicit (drawn)	**Implicit (word)**
Lion	Run
Flower	Smell
Bee	Stay still/away
Bed	Go to sleep
Fire truck	Pull over in the car
Stop light	Depends on the color—Stop/slow/go

Friend	Hug
Enemy	Stay away/avoid
Baby	Coo
Puppy	Pet
Flag	Salute/hand over heart

Memory Challenge

With a partner, each person should examine one tray of objects. They should look for a few minutes. Then remove the tray. Ask them if they can remember what was on the tray.

1. Tray #1: First-Aid Objects
   Bandages
   Neosporin antiseptic
   Small scissors
   Tweezers
   Cotton ball
2. Tray #2: Writing Objects
   Pencil
   Paper
   Wite-Out
   Eraser
   Paper clips
3. Tray #3: Camping Objects
   Jackknife
   Small rope
   Candle
   Insect repellent
   Matches
4. Tray #4: Hair Care Objects
   Rollers
   Hair spray
   Comb
   Brush
   Barrette

Now, switch trays. Can they still remember?

One final time, switch the trays and exchange what is on the trays. For example, mix up camping objects with hair care. Now can they remember what was on the first tray?

Our brains are more skilled at remembering pictures than words. During the first two times the objects were in a linked series (Tray #1 first aid, Tray #2 writing, etc.). The final time, the objects were all a jumble, thus making it more difficult for the brain to recall. Tell the children to try this at home with their parents to see how they do!

Garden Search

1. Send the pairs out into the garden together to identify three things on a piece of paper: a plant, a plant container, and a work of art.
2. Tell the group to list these three things on a sheet of paper then return to the larger group and give the paper to the leader for further instructions.
3. Complete the Brain Diagram (at the end of this section) by coloring in the brain as labeled and identifying the different areas of the brain with colored pencils.
4. After completion of the Brain Diagram, ask partners to identify the three things they chose in the garden. Ask them, "How long did your long-term memory work in this activity?"

Evaluation

At the end of the session, ask participants to write the answers to the following questions in their journals:

1. What is the limbic system?
2. What are the implicit and explicit systems in the limbic?

Reinforce and wrap up by asking participants to thank one another for their participation and working together so well.

Definitions

brainstem: The brainstem is another small part of the brain that is mighty in its function. The brainstem sits beneath the cerebrum and in front of the cerebellum. It connects the rest of the brain to the spinal cord, which runs down the neck and back. The brainstem is in charge of all the functions our bodies need to stay alive, for example, breathing, digesting food, and circulating the blood in our veins.

cerebellum: The cerebellum (pronounced "sair-uh-beh-lum") is at the back of the brain, below the cerebrum. It is a small part of the brain doing big work. It is in charge of controlling balance, movement, and coordination of muscles. Thanks to the cerebellum, we can stand up, sit up straight, and move around.

cerebrum: The cerebrum (pronounced "sur-ee-brum") is the largest part of the brain. It makes up 85 percent of the brain's weight. It is the thinking part of the brain. It lets us solve math problems, play games, feed the dog, remember our favorite flowers, and create art. The cerebrum is the part of the brain that lets us reason. It is made up of two halves, with one on either side of the head—the left brain and the right brain. Some scientists believe the right brain is the abstract thinking part of the brain; it helps us think about music, colors, and shapes. The left brain is believed to be more analytical; it helps us with math, speech, and logic. Scientists do not know for sure if the right half of the cerebrum controls the left side of your body and the left half controls the right side.

hypothalamus: The hypothalamus (pronounced "hy-poh-tha-luh-muss") sits right in the center of the brain and is responsible for the temperature of your body. It knows to keep the body's temperature at about 98.6°F or 37°C. It is the part of the brain that sends messages to the body to either sweat or shiver: sweat if we are too hot; shiver if we are too cold.

thalamus: The thalamus (pronounced "tha-la-mus") is a mass of nerve cells centrally located in the brain just below the cerebrum and resembling a large egg in size and shape. The thalamus is a routing station for all incoming sensory impulses. In addition, it connects various brain centers with others. Thus the thalamus is a major integrative complex, enabling sensory stimuli to summon appropriate physical reactions as well as to affect emotions. Working with the hypothalamus, the thalamus establishes levels of sleep and wakefulness. It is also vital to the neural feedback system controlling brain wave rhythms.

Week #2: The Brain Diagram

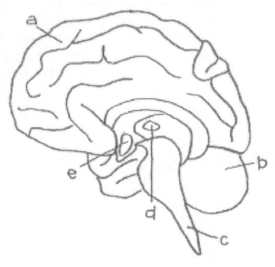

a. Cerebrum . . . . Gray
b. Cerebellum . . . . Pink
c. Brainstem . . . . Green
d. Thalamus . . . . Yellow
e. Hypothalamus. . . . . . . Purple

WEEK #3: SOUND AND THE EAR

General Statement of Purpose

The purpose of this activity is to introduce participants to the sense of sound. It is also the purpose of this activity to introduce the participants to the world of the honeybee.

Goals

1. To teach elementary introduction to sound sense
2. To teach basic entomology of honeybees
3. To encourage social interaction between elders and children

Materials and Equipment

- Multiple copies of Ear Diagram (end of section)
- Enlarged honeybee photos (see Resources)
- Large pieces of paper or chalkboard
- Masking tape
- Magnifying boxes filled with different types of bees (wasps, yellow jackets, hornets, bumble bees, mason bees, honeybees)
- Three different kinds of honey (e.g., clover, basswood, and buckwheat)
- Toothpicks
- Three small bowls
- Multiple copies of "Honeybee Poem"
- Notebook journals
- Variety of flowers (pansy, delphinium, iris) that depict how the bee sees the blossom of a flower as a "landing pad"

Program Procedures

Introduction to Sound—Use Ear Diagram to answer the following questions.

What Is Sound?: A Definition

The processes of hearing and balance involve the stimulation of hair receptor cells in the inner ear. Sounds are created by waves of pressure that cause the air to vibrate. These vibrations trigger a chain of movement from the outer ear to the inner ear, where hair cells send electrical impulses to the brain for inspection.

Questions to Consider

Ask participants to consider the following questions.

1. Did you know that humans can distinguish more than 1,500 musical tones?
2. Did you know people can hear sounds ranging from 0 to 140 decibels? Now that is quite a range! Consider that 0 decibels is no noise at all and 140 decibels is a plane at 10 yards.
3. Did you know that ears can detect the direction of sound within 3 degrees?
4. Did you know the smallest bone in the body is the stirrup bone found in the middle ear?

Honeybees

Introduce the group to this week's lesson:
"Today we are going to learn about ears in their relationship to listening to an insect. That insect is the honeybee. Do you know the name of the insect family honeybees come from?" Write down the word *Hymenoptera*. List members of the family of bees:

Honeybees
Bumble bees
Wasps/yellow jackets/hornets
Ants

Note that all bees do not look and act alike. Give examples of differences in size and behavior, for example, wasps and ants have shiny, hard bodies while bumblebees and honeybees are fuzzy and furry.

Use enlarged photos of honeybees to explain the following points. See Resources to order photos from Dadant and Sons Beekeeping Company.

1. What makes a honeybee different from the other bees mentioned?

   • Honeybees make honey.
   • They spread pollen.
   • They live in colonies maintained by people.

2. Who are honeybees?

   • Workers
   • Drones
   • Queen

3. What do the workers do in the colony?

   • They build, repair, and keep clean the honeycomb.
   • They take care the young brood.
   • They go into the fields to gather nectar and pollen.
   • They maintain the hive at just the right temperature (95°F).
   • They look after the queen.
   • They guard the hive from invaders.
   • They are the investigators of the hive.
   • Their feet "taste" what they touch.

4. What does the queen do for the hive?

   • The queen lays eggs (up to 100,000).
   • She is the reason for the hive.

5. What is the drone?

   • The drone is a male bee who mates with the queen.

6. What do bees see?

- Bees see the full spectrum of ultraviolet color.
- They see targets and landing pads on flowers (show examples of pansy, delphinium, etc.) as they fly high above the plants.

7. How many flowers does a worker bee visit to make one pint of honey?

- Five million flowers.

8. How do bees make honey?

- Bees drink nectar and then store it in their "honey crop" stomach. Bees have two stomachs—one to store nectar, the other to store its own food for resources in flight and out in the field. The bees bring the nectar back to the hive in their honey crop and then regurgitate the nectar into a cell. From there, the liquid evaporates by the fanning of the bee wings, and the bees store the honey until it is thickened. The top of the honey is capped to keep it from drying out and getting hard. The beeswax is secreted from a special gland in their abdomen. One ounce of wax is equal to one pound of honey. In other words, it takes one pound of honey for the bee to produce one ounce of wax.

9. How do bees talk to each other?

- Bees dance in the shape of the letter S, circling either toward or away from the sun—depending on the location of the nectar source.
- They smell through their pheromones—that is, the workers touch the queen constantly to smell her pheromones and then they pass her smell from their antennae to their mouthpart and pass it on to others in same fashion. This is all to assure the hive that everything is okay, and it keeps the bees in the colony and the queen laying eggs.

Activities

Honey Testing

Each member tastes the three different honeys (use toothpicks). Ask them which kind they like best.

Honeybee Poem

Ask the group to read the "Honeybee Poem" (found at end of section) as partners, then as a large group. One person takes the left column, and the other takes the right column. As a large group, half of the group takes the left column, and the other takes the right column.

Hearing Activity Page

Color in the Ear Diagram (end of section) and identify the parts with your partner.

Evaluation

Ask individuals to write in their journals the answers to the following questions:

1. Who are honeybees?
2. What are their different roles?
3. What sound does the honeybee make?
4. What did you like best about working with your partner today?
5. What did you learn today that you will never forget?

Thank them for their participation. Also remind them to say thanks to the honeybee whenever they see them at work in the garden.

Definitions

anvil: The malleus or hammer is attached to another tiny bone called the incus, which means "anvil" in Latin.

auditory canal: This is the name for the ear canal. It is the passage through which sound waves travel toward the middle ear.

auditory nerves: As the cochlea move, sounds vibrations are transformed into nerve signals, the language of the brain. Once sound becomes nerve signals, the brain can understand what the ears are hearing. The brain puts it all together. The brain would not be able to recognize any sounds without the different parts of the ear all working together.

cochlea: The vibrations entering the inner ear go into the cochlea, which is a small, curled tube in the inner ear. The cochlea is filled with liquid and lined with cells that have thousands of tiny hairs on their surface. These hairs are not like the ones on your head; these hairs are so small that they can only be seen with a microscope. When sound vibrations hit the liquid in the cochlea, the liquid begins to vibrate. Different kinds of sounds will make different patterns of vibrations. The vibrations then cause the tiny hairs in the cells of the cochlea to move. The more vibrations, the more the hairs in the cochlea will move.

eustachian tube: The eustachian tube is involved in the function of the middle ear. The middle ear is connected to the back of your nose by a narrow tube called the eustachian tube. Together the middle ear and the eustachian tube keep equal air pressure on both sides of your eardrums. The change in pressure can sometimes feel like a "pop" after an airplane ride or after you have yawned. This feeling of is the eustachian tube opening to try to equalize air pressure on both sides of your eardrum.

hammer: The tympanic membrane is a piece of thin skin stretched tight like a drum. It is attached to a small bone called the *malleus* which means "hammer" in Latin.

outer ear: This is the part of the ear that people can see. It is called the pinnea or auricle. Part of what the outer ear does is protect the ear by making wax. This special wax contains chemicals that fight off infections which could hurt the skin inside the ear canal. The wax also helps keep dirt from the ear. However, the main job of the outer ear is to collect sounds. When sound occurs, sound waves are produced and travel through the air. After these waves enter the outer ear, they travel through to the ear canal.

semicircular canals: These canals are the part of the ear that helps establish our sense of balance. These small loops are located right above the cochlea. They are filled with liquid and have thousands of microscopic hairs that send messages to the brain for balance. When you move, the liquid in the semicircular canals moves too.

tympanic membrane: The tympanic membrane is within the middle ear. The middle ear takes the sound waves it receives from the outer ear, turns them into vibrations, and ultimately delivers them to the inner ear. It does all this using the eardrum or technically known as the tympanic membrane. The tympanic membrane actually separates the outer ear from the middle ear.

Week #3: Honeybee Poem

Instructions: Read this poem together as a group and in two parts. When words are in both columns, it should be read as a group. When words are in only one column, the other half of the group is silent.

1. Buzz	1. Buzz
2. (silence)	2. I like being a queen bee.
3. We are the worker bees.	3. (silence)
4. We will gladly explain.	4. I will gladly explain.
5. (silence)	5. All day long I am fed and fanned by my royal attendants.
6. We are up at dawn guarding the hive.	6. (silence)
7. (silence)	7. I am fed and bathed.
8. We are out in the garden gathering pollen and nectar.	8. (silence)
9. (silence)	9. I am groomed and kept at a constant temperature.
10. We are inside making wax without any time to relax.	10. (silence)
11. (silence)	11. I lay eggs by the thousands.
12. We take care of the larva, feeding the little brood.	12. (silence)
13. (silence)	13. I am forever loved and applauded.
14. We are still packing combs with the pollen and making honey for the thousands.	14. (silence)
15. (silence)	15. I relax for a moment.
16. Although we are weary we keep patching cracks in the hive with our propolis.	16. (silence)

17. (silence)	17. Then I retire for the day.
18. We still work and build, making the hive suitable for our queen.	18. (silence)
19. Truly a bee's life is the best of them all.	19. Truly a queen's life is the best of them all.

Source: Inspired by Paul Fleischmann, *Joyful Noise: Poems for Two Voices,* by Harper Row, 1988.

Week #3: Ear Diagram

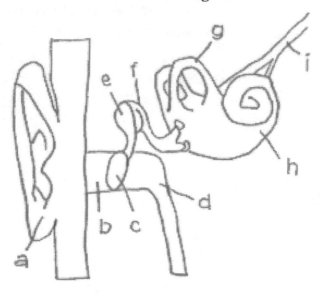

a. Outer ear . . . . Tan
b. Auditory canal . . . . Pink
c. Tympanic membrane . . . . Red
d. Eustachian tube . . . . Orange
e. Hammer. . . . . Light brown
f. Anvil . . . . Green
g. Semicircular canals . . . . Yellow
h. Cochlea . . . . Blue
i. Auditory nerves . . . . Purple

WEEK #4: SIGHT AND THE EYE

General Statement of Purpose

The purpose of this activity is to introduce participants to the sense of sight.

Goals

1. To learn how the brain works in relationship to the sense of sight
2. To teach basic horticultural science
3. To encourage social interaction between elders and children

Materials and Equipment

- Multiple copies of the Eye Diagram
- Drawing paper
- Colored pencils
- Pencils
- Blindfold
- Notebook journals
- Multiple copies of Find Flowers chart

Program Procedures: Introduction to Sight

What Is Sight?

Use the Eye Diagram (end of section) to explain the following:

"The eye changes light into signals that can travel along nerves to the brain. At the front of the eye is the cornea, a transparent layer which bulges slightly so it can bend the light that enters the eye. Once it has passed through the cornea, the light travels through the pupil, the lens, and then a transparent jelly (called the vitreous humor) before it finally reaches the retina. The lens bulges far more than the cornea. Together, the lens and cornea focus an upside-down image of the outside world on the retina—or the membrane that lines the eye."

Activities

Sculpture

Ask participants to form groups of three. This activity requires members to play three roles:

1. *The model*—This person will stand or sit in a comfortable position that he or she chooses.
2. *The sculptor*—This person will wear the blindfold and will be given as much time needed to discover, through touch and no sight, what position the model has chosen. Then he or she will sculpt the clay into that position.
3. *The clay*—The clay is the person who will be sculpted into the same position that the model has chosen.

As time allows, ask members to switch roles. Ask participants the following questions and have them write answers in their journals:

How did it feel to be the model?
How did it feel to be the sculptor?
How did it feel to be the clay?
Did you regret not being able to see?

Partner See, Partner Draw

Participants should be in pairs. This activity requires one person to look at something in the room and then describe it to his or her partner. The partner cannot look at the object. He or she must draw it based on the description. The person looking at the object cannot tell his or her partner what the object is; the person can only describe it. As time allows, switch roles. Ask the participants the following questions to answer in journals:

How did it feel to draw something that you could not see?
How did it feel to only be able to describe the object and not tell your partner to look at it or identify what it is?
Did you regret not being able to see?

Find Flowers

With the Find Flowers chart, ask partners to go into the garden to find the flowers that correspond to the colors on the chart (end of section). Identify any insects on the flower. As a large group, ask them to share their findings.

Evaluation

Ask partners to color in the Eye Diagram (end of section) and identify the parts of the eye together.

Afterward, ask them to answer the following questions in their journals:

1. How would you feel if you could not see at all?
2. How will you be able to understand and appreciate others who cannot see now that you have done these exercises?
3. What did you learn today about sight that you will never forget?

Thank them for their participation and working with their partners.

Definitions

cornea: The cornea is the window to view the world through. It is part of the sclera in front of the colored part of the eye. The cornea is transparent and lets light travel through it. The cornea helps the eye focus as light makes its way through. It is a very important part of the eye, however, you can hardly see it because it is made of clear tissue.

iris: The iris is the colorful part of the eye. When a person has blue eyes, it really means he or she has blue irises. The iris is attached to muscles that change shape. This allows the iris to control how much light goes through the pupil.

lens: After light enters the pupil, it hits the lens. The lens sits behind the iris and is clear and colorless. The job of the lens is to focus light rays on the back of the eyeball—much as the lens of a movie projector focuses the images onto the film screen.

optic nerve: The optic nerve is the messenger to the brain. The retina changes the images you see into millions of nerve messages. Then the optic nerve carries those messages from the eye to the brain. The optic nerve serves as a high-speed "telephone line" connecting the eye to the brain. When you see an image, your eye "calls" your brain with a report on what you are seeing so the brain can translate that report into "flower," "tree," or whatever the case may be.

pupil: Behind the cornea is the pupil. The pupil is the black circle in the center of the eye. It lets light enter the eye.

retina: The retina is in the very back of the eye. The retina is where the lens focuses the light rays. Though the retina is smaller than a dime, it holds millions of cells that are sensitive to light. The retina takes the light the eye receives and changes it into nerve signals so the brain can understand what the eye is seeing.

sclera: The sclera is the white part of the eyeball. It is made of a tough material and has the important job of covering most of the eyeball. Think of the sclera as your eyeball's outer coating. If you look very closely at the white of the eye, you will see tiny lines that look like pink threads. These are the blood vessels that deliver the blood to the sclera.

vitreous humor: The vitreous humor is the biggest part of the eye and sits behind the lens. The vitreous body forms two-thirds of the eye's volume and gives the eye its shape. It is filled with a clear, jelly-like substance.

Week #4: Eye Diagram

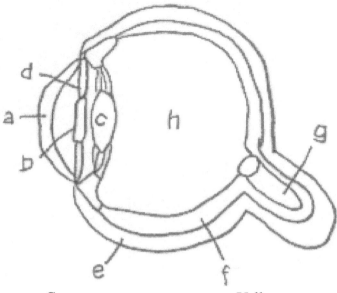

a. Cornea . . . . Yellow
b. Pupil . . . . Gray
c. Lens . . . . Pink
d. Iris . . . . Blue
e. Sclera . . . . White
f. Retina . . . . Green
g. Optic nerve . . . . Red
h. Vitreous humor . . . . Orange

Week #4: Find Flowers Chart

Flower Colors

Flower Colors	Flower Visitors
Blue	🐝
Purple	🐝
Pink	🐝
Orange	🐝
Yellow	🐝
White	🐝
Red	🐝

Flower Visitors

WEEK #5: THE SENSE OF TOUCH

General Statement of Purpose

The purpose of this activity is to introduce participants to the sense of touch and tactile stimulation.

Goals

1. To recognize the sense of touch
2. To teach basic horticulture science
3. To encourage social interaction

Materials and Equipment

- Toothpicks
- Multiple copies of Skin Diagram (end of section)
- Unfrosted sugar cookies (one per person)
- Cans of colored decorating frosting
- Napkins
- Blindfold
- Notebook journals
- Multiple copies of Flower Shapes (end of section)
- Pencils

Program Procedures: Introduction to Touch

What Is Touch?: Points to Share

The skin is the waterproofing barrier that protects the body from physical damage and infection. It is also sensitive to touch helps control body temperature, and repairs itself.

Read the following to the participants:

Did you know these facts? (Use Skin Diagram at end of section.)

1. The skin is the largest organ of the body.
2. The skin has a surface area of up to twenty-two square feet.

3. The body sheds over forty pounds of skin in an average lifetime.
4. Household dust is mainly dead skin cells.
5. The skin has two main layers: an outer *epidermis* and an inner *dermis.*
6. New cells move from the base of the epidermis to the surface, where they harden and die, producing a tough waterproof layer. The dermis is living and contains nerves, blood vessels, sense receptors, glands, and hair follicles.
7. The skin contains a variety of nerve endings that detect light, touch, sustained pressure, cold, heat, and pain. They send electrical signals to the brain. When the tissue is damaged, cells release chemicals. These chemicals activate the bare nerve endings that detect pain.

Activities

Two-Point Test

Give each set of partners a few toothpicks. Have one partner perform the action and the other receive the action. Read these directions to the participants:

1. Close your eyes and have your partner touch you lightly with the toothpick at two different points on your body (e.g., on your arm and on your foot).
2. Let your partner know where you felt the toothpick touching you, and what part had more feeling in it when the toothpick touched.
3. Have your partner record your results and repeat the test , five more times touching two more points each time. As times allows, switch roles.
4. Do you find that some areas of your body are more sensitive than others?

Words on a Back

Again, one partner acts, the other receives in this activity. Read these directions aloud:

1. Close your eyes. Have your partner touch you lightly on the back, while "writing" a word. He or she will use only a fingertip to write this word.
2. Let your partner know what you are feeling as he or she writes. Can you identify what the word is? If so, share your guess with your partner. If you are right, trade roles. If not, continue until you guess the correct word.

Flower Shapes

One partner should be blindfolded. The other should lead the pair to the garden to find and feel flowers with different shapes as listed on the chart (see end of this section for flower shape chart). Ask the blindfolded participant these questions and have them answer in their journals. Switch places.

What are some of the features of the flower shapes?
Does the plant allow a place for the insect to stand?
Is there a tube for the insect to drink the nectar?
What else can you feel on the flower?

Ask each pair to share their findings with the group.

Flower Cookies

Provide cookies, frosting, and napkins to each pair. Ask each person, with his or her partner, to re-create the shape of one of the more intricate flowers found in the Flower Shapes exercise. They will use frosting as the paint and the cookies as the canvas. Each pair should show their creations to the group. Then all participants should enjoy their cookie art, eating together.

Evaluation

Wrap-Up

Review the Skin Diagram. Thank the group members for their participation and working with their partners.

Definitions

blood vessels: The dermis is full of blood vessels. These keep your skin cells healthy by bringing them the oxygen and nutrients they need. Blood vessels also carry away waste. Sometimes it's hard to look at our own blood vessels. But take a look at your elder partner; as a person ages—the dermis gets thinner, and the skin becomes more transparent. Can you see the blood vessels in your elder partner's hand?

dermis: The dermis (pronounced dur-miss) is the next layer after the epidermis. You cannot even see the dermis because it is hidden underneath your epidermis. Your dermis contains the nerve endings, blood vessels, oil glands, and sweat glands. All of these parts play a key role in keeping your skin healthy.

epidermis: The epidermis (pronounced eh-pih-dur-miss) is the outside layer of skin, the part of your skin you can see. Your epidermis is hard at work making new skin cells all the time. Right now, you are shedding old skin at about 30,000 to 40,000 dead skin cells every minute of the day! That is nearly ten pounds of skin cells a year. But never fear—your skin is replacing the old with new all the time. In fact, 95 percent of cells in the epidermis are working to make new skin.

hair: Your hair follicles rely on the subcutaneous glands to bring on the shine. Connected to each follicle in the dermis layer is a tiny sebaceous gland that releases sebum onto the hair. This lightly coats the hair with oil, giving it shine and a little waterproofing.

hair follicle: The subcutaneous layer holds the root of the hair follicle. Each hair on your body grows out of a tiny tube in the skin called a follicle. Every follicle begins in the subcutaneous layer and continues up through the dermis. You have hair follicles all over your body except on your lips, palms of your hands, and soles of your feet.

nerve endings: These are responsible for telling your brain how things feel when you touch them. They work with your nervous system in your brain to let your hands know when things are hot or cold, rough or smooth. Nerve endings are all over your body—in your lips when you eat cotton candy and in your legs when a mosquito bites you. The nerve endings work with your muscles to keep you from getting hurt. If you touch something that is too hot, the nerve endings in your dermis respond right away. The nerves send this message quickly to your brain and spinal cord, which then immediately tells the muscles to take your hand away from the hot object. This happens in less than a second, thanks to the nerve endings in your dermis!

pore: Sweat comes out on to the epidermis through pores. Pores are tiny holes in the skin that allow sweat to escape. When the sebum meets the sweat, they form a protective film that can feel sticky.

sebaceous (oil) glands: The sebaceous glands (pronounced: she-bay-shuss) are always producing sebum. Sebum is a natural oil produced in your skin. The sebum rises to the surface of the epidermis to keep your skin lubricated and protected. It also makes your skin waterproof—as long as sebum is on the skin, your skin won't get soggy.

subcutaneous layer: The subcutaneous (pronounced: sub-cyoo-tay-nee-us) layer is the third and bottom layer of skin. It is made up mostly of fat and helps your body stay warm and absorb shocks. The subcutaneous layer also helps bind your skin to all the tissues underneath it.

sweat gland: Sweat is your body's way of cooling itself. When you sweat, your sweat glands excrete sweat through your epidermis. Even though you cannot feel it, your glands are producing sweat all the time. Sweat is essential for your body to stay cool and to maintain our 98° temperature.

Week #5: Skin Diagram

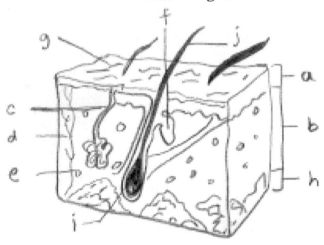

a. Epidermis . . . . Yellow
b. Dermis . . . . Orange
c. Sweat gland . . . . Blue
d. Nerve ending . . . . Red
e. Blood vessels . . . . White
f. Sebaceous (oil) gland . . . . Purple
g. Pore . . . . Black
h. Subcutaneous layer . . . . Pink
i. Hair follicle . . . . Black
j. Hair . . . . Black

Week #5: Flower Shapes Chart

Flower shapes that bring butterflies

little tubes to drink from	a place to stand
	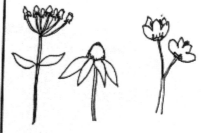

WEEK #6: THE SENSES OF TASTE AND SMELL

General Statement of Purpose

The purpose of this activity is to introduce the science of the senses of taste and smell.

Goals

1. To encourage social interaction
2. To teach the basis of taste and smell
3. To learn basic horticulture science

Materials and Equipment

- Multiple copies of Tongue Diagram (end of section)
- Multiple copies of Nose Diagram (end of section)
- Blindfolds
- Lemon juice
- Unsweetened chocolate
- Honey
- Soy sauce
- Four small bowls
- Toothpicks
- Large salad bowl
- Bagged salad greens
- Salad dressing
- Edible herbs and flowers
- Unfrosted sugar cookies
- Cans of vanilla frosting

Program Procedures

What Is Taste? What Is Smell?

Introduction to taste and smell—Use Tongue Diagram and Nose Diagram to cover the following points:

1. The tongue is used in talking, eating, and tasting.
2. The nose functions to identifying smells.
3. The tongue is a large muscular structure. It contains taste buds that detect chemicals in food and drink. The tongue works with the nose to identify foods.
4. The nose's fine hair like structures (cilia) inside the nasal cavity process the airborne molecules, which dissolve in the nasal mucus and stimulate strong emotional responses in the brain. The cilia cover the olfactory bulb that are sent to chemicals in the mucus. The olfactory nerve sends messages to the brain.

Questions to Consider

Ask group members these questions:

1. Did you know that taste is 75 percent smell?
2. Did you know that a moth can smell a single molecule of another moth's pheromone?
3. Did you know that everyone has a unique smell, except identical twins?
4. Did you know that dogs can distinguish nonidentical twins by smell—but not identical twins?
5. Did you know that insect antennae when attached to electronic circuits can be used as odor sensors?
6. Did you know we can smell happiness?
7. Did you know that some people cannot smell a skunk?
8. Did you know a bloodhound can pick up a twenty-four-hour-old trail and identify the person?
9. Did you know a baby has taste buds all over the inside of its mouth until it is three months old?
10. Did you know there are more than 10,000 taste buds mainly on the tongue?
11. Did you know taste bud cells only last a week before they are renewed?
12. Did you know the sense of smell can detect more than 4,000 different odors?
13. Did you know the brain grows used to smells and quickly stops registering them?

14. Did you know that the nose and the tongue work together for taste? The taste buds (receptor cells) are located on the tongue's surface. They are called papillae. The olfactory bulb takes in the molecules that are detected in the olfactory bulb. The tongue and olfactory bulb determine the detection of the four basic flavors—sweet, sour, bitter, and salty.

15. Did you know that bitterness is located at the back of the tongue? Sour is on either side of the tongue. Sweet is found at the tip of the tongue. Salty lies directly along either side of sweet.

Activities

Tongue and Nose Diagrams

Each pair should color in the Tongue and Nose diagrams and identify parts on sheets (end of section).

Taste Test

Break group into pairs. One partner, will receive the action and the other will perform the action.

With the blindfold on one partner, the other partner dips a toothpick into four different substances. The blindfolded partner tastes and identifies whether the substance is sour, bitter, sweet, or salty. As time allows, switch roles. Use the following items:

Lemon juice (sour)
Unsweetened chocolate (bitter)
Honey (sweet)
Soy sauce (salty)

Edible Flower Salad

Each pair should go into the garden and find flowers and herbs that are edible, such as pansies, nasturtiums, roses, herbs (thyme, basil, parsley, oregano, rosemary, chives) marigolds, lavender, and day lilies. Ask them to bring ingredients back to the larger group.

The group should wash and dry the plants and then prepare a salad (add a bag of prewashed salad greens to the flowers). Salad dressing should be added, then the whole group can eat salad together.

Edible Flower Cookies

Again, each pair should go into the garden and find edible flowers and herbs. Ask them to bring the ingredients back to the large group. Again, wash and dry the plants. Each pair should frost large cookies. Then they will decorate their cookies with real flowers, and eat them.

Evaluation

After the group is finished enjoying their cookies, ask them to clean up the area and answer the following questions in their journals:

1. What are the four flavors?
2. What is the olfactory bulb?
3. What is your favorite edible flower?
4. What did you enjoy most with your partner today?

Thank them for their participation in cleaning up and working with partners.

Tongue Definitions

papillae: The layer of bumps on your tongue are called papillae. The purpose of the papillae are to grip food and move it around while you chew. The papillae are divided into three different types—two in front and one in the back of the tongue. At the front of the tongue are the fungiform papillae and filiform papillae. The larger ones are the fungiform papillae, and the smaller ones that look a little like hair are the filiform papillae. At the back of the tongue are the vallate papillae. They are the large and round, and there are about eight to twelve of them.

saliva: Salive helps break down food. Saliva helps food travel on the tongue, down the throat, and into the stomach.

salivary glands: These glands produce saliva.

taste buds: The taste buds are a part of the papillae. The taste buds allow you to taste food flavors. You are born with about 10,000 taste buds. However, as you age, your taste buds die. The taste buds detect sweet, sour, bitter, and salty foods. The taste bud has a microscopic hair called microvilli. The microvilli are covered with taste receptors. When you put something in your mouth and it begins to be digested by the saliva, this stimulates the microvilli to start making nerve signals. These signals then travel to the brain, where the brain can interpret the signals and identify the taste for you.

Nose Definitions

cilia: The cilia are microscopic hairs that are covered with special receptors which are sensitive to odor molecules that travel through the air. These receptors are very small—there are at least 10 million of them in your nose. There are at least twenty different kinds of receptors, and each kind has the ability to sense a certain range of odor molecules.

olfactory bulb: This is the spot right underneath the front of your brain at the top of the nasal cavity. The brain's job is to interpret the nerve signals and identify the smell for you carried up from the olfactory bulb. Identifying smells is the brain's way of telling you about your environment and keeping you safe. Think back to the last time you smelled a cat's dirty litter box: In an instant, your olfactory epithelium and olfactory nerve worked together to get the message to your olfactory bulb. Once your brain unscrambled the nerve impulses, it recognized the smell as "icky" and you knew it was time to clean the litter box.

olfactory epithelium: The olfactory epithelium is on the roof of the nasal cavity, or the hole behind your nose. The olfactory epithelium is the tiny patch of nerve cells that houses the microscopic hairs known as cilia.

olfactory nerves: When odor enters your nose, it stimulates the cilia to start producing nerve signals. The nerve signals move along the receptors and travel to the olfactory nerve, which then transmits the signals to the olfactory bulb.

Week #6: Tongue Diagram

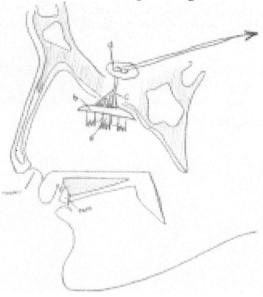

a. Papillae . . . . Red
b. Taste bud . . . . Orange
c. Salivary glands . . . . . Purple
d. Saliva . . . . Blue

Week #6: Nose Diagram

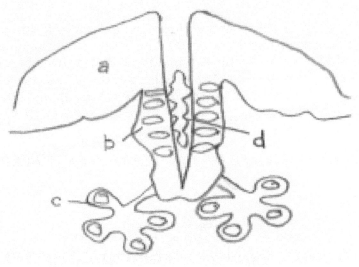

a. Cilia. . . . . Yellow
b. Olfactory epithelium . . . . . Gray
c. Olfactory nerves. . . . . Pink
d. Olfactory bulb. . . . . Blue

POSTPROGRAM PARTY

The group of elders and children should return to the project center for a postprogram luncheon together. A picnic lunch should be provided. Group members should be asked to share their favorite stories from the summer. Each member should be given a certificate of participation (see example). A group picture might be taken. Project participants visit more and say good-bye.

CERTIFICATE OF COMPLETION EXAMPLE

Congratulations!
You have completed the Sojourn Summer 2001
Sense-Ational Garden program
at the University of Minnesota Landscape Arboretum

Signed:

Jean M. Larson
Coordinator of Therapeutic Horticulture

Chapter 5

Evaluation Strategies

RATIONALE TO EVALUATION

Evaluating intergenerational programs is important for the following reasons:

- It provides an actual measurement of how the program has benefited the participants.
- It provides inspiration for future programs.
- It provides information to determine which projects and activities to continue and which to discontinue.
- It provides an opportunity for participants to have a voice in the program process, that is, how it is facilitated and what the program contains.

EVALUATION EXAMPLE

The main goal of the intergenerational gardening program is to provide an opportunity for participants to socially interact. Therefore, the evaluation form should reflect this goal in its content questions. Below is an example of an evaluation the elders and children were asked to fill out on a session-by-session basis.

1. On a scale from 1 to 7 (see scale below), how satisfied were you with the overall intergenerational garden program today? Please circle your answer.

doi:10.1300/5531_05

1 Extremely dissatisfied
2 Very dissatisfied
3 Somewhat dissatisfied
4 Dissatisfied
5 Moderately satisfied
6 Very satisfied
7 Exceeded all expectations

2. Please describe how working with your partner makes you feel. Why?
3. Please make a list of the things you enjoyed most about working with your partner during the program today.
4. Is there anything you did not like about working with your partner in the program today? Please tell us.
5. Would you come to this program again in the future? Yes or no (Please circle one.) Why or why not?
6. Do you have suggestions on topics, improvement, and/or other comments regarding the intergenerational garden program?

POSTEVALUATION

Once you have collected the evaluation information, you have a chance to track the data to see if there are any trends or indicators for future program development. For example, you could place numerical values on the responses the participants give to each of the weekly activities. Then you can add them up to compare them for each participant from beginning to end. You could use your findings to adjust future programs. You could also use the information to report to the sponsoring agency the success of the program as support for allocating more funds for future programs. However the data is used, it is an important aspect of the program because it provides the summary language for present and future. Here are a few examples of comments made by participating elders and children:

Children's Comments

- "I love working with Joe—he is so funny and nice."
- "She is just like my grandma. I miss my grandma—she died last year."
- "We went to the garden and found lots of honeybees on the catnip—it was *cool!*"
- "I will come back next year for sure! I hope that Bob comes back, too. I liked him a lot."

Elders' Comments

- "She is such a sweet little girl, remembered all the names of the flowers."
- "Andrew is a very thoughtful young man. He has been most helpful pushing my wheelchair in the garden."
- "It's dirty work (being in the garden), but it sure beats sitting around at home by myself."
- "Playing in the garden with all of these children makes me feel young again. Yeah, like I'm a kid playing with dirt."
- "Clare and I will see each other next summer and sooner than that. We found out we go to the same church."

Resources

BASIC HORTICULTURE

Over the years, through success and failure, we've had to access books, Web sites, and other professional material for wisdom and advice. Following are a few of our favorites compiled in a selected resource list.

John Jeavons has been one of our personal gardening heroes for a long time. He brings the idea of planting many things together in a small space. This is ideal for those in health care who have a very small garden area in which to carry out their programs. This is a must-have for any educational/treatment library.

Jeavons, John (1995). *How to Grow More Vegetables.* San Fransico: Ten Speed Press. I love the photos in this book.
Kite, L. (2001). *KISS (Keep It Simple Stupid) Guide to Gardening.* New York: Dorling Kindersley Publishing, Inc.

BOTANIC GARDENS AND ARBORETA

Here is a sample listing of a few botanical gardens that feature a program in therapeutic horticulture. Visit the American Horticultural Therapy Association (AHTA) Web site at www.ahta.org for more botanic gardens in your area.

Brooklyn Botanic Garden
www.bbg.org

Cheyenne Botanic Gardens
www.botanic.org

doi:10.1300/5531_06

Chicago Botanic Garden
www.chicago-botanic.org

Cleveland Botanical Garden
www.cbgarden.org

Denver Botanic Gardens
www.botanicgardens.org

Guelph Enabling Garden
http://www.enablinggarden.org/

Minnesota Landscape Arboretum Clotilde Irvine
http://www.arboretum.umn.edu

The New York Botanical Garden
www.nybg.org

HONEYBEES

Honeybees are one of our favorite subjects. Many nonfiction books on the market now are a great read. The Dadant and Sons Beekeeping and Beekeeping Supplies Web site can help you with some resources that are more educational in their presentation.

Dadant and Sons Beekeeping and Beekeeping Supplies
www.dadant.com

At this site, many prints and books can be ordered. See the on-line catalog and section under magazines, books, and videos for many resources, including the following:

Honeybee Life Cycle Chart
Honeybee Study Prints
The Hive and the Honeybee (book)

HORTICULTURAL THERAPY

Here is a listing of a few therapeutic programs found throughout the United States and Canada. Again, visit the AHTA Web site at www.ahta.org for a complete listing.

Canadian Horticultural Therapy Association
http://www.chta.ca

City Farmer
www.cityfarmer.org/horttherp70.html#hort

Enid A. Haupt Glass Garden/Rusk Institute
http://www.med.nyu.edu/rusk/glassgardens/index.html

Horticultural Therapy Program at Kansas State University
www.oznet.ksu.edu/horttherapy/

Meristem-Restorative Landscapes For Health Care Environments
www.meristem.org

Thrive
www.thrive.org.uk/

Virginia Tech, Department of Horticulture, Horticultural Therapy information
www.hort.vt.edu/human/human.html

INTERGENERATIONAL PROGRAMS

While this sourcebook covers intergenerational gardening, many Web sites discuss serving the needs of elders and children together. Here are just a few.

Intergenerational Innovations—An intergenerational program in Seattle, Washington
www.intergenerate.org

Generations Together—Pittsburgh University Intergenerational Studies Program
www.gt.putt.edu/

Generations United—National organization to promote intergenerational strategies, programs, and policies
www.gu.org

Gray Panthers—Organization
www.graypanthers.org

The Generations of Hope—An alternative retirement community and adoptive service based on generations working together to raise "unadoptable" children
www.generationsofhope.org

Temple University—The Center for Intergenerational Studies
www.temple.edu/cil

SENSORY GARDENS

Diane Akerman will take you into the world of the senses through her narrative of plants and animals. This is a wonderful book to quote from in programming.

Akerman, Diane (1990). *A Natural History of the Senses.* New York: Vintage Books.

The following Web site offers insight into how gardens can be a sensory experience for everyone. The Sensory Trust in Bath, England, is a resource for professionals in the United Kingdom as well as the rest of the world.

www.sensorytrust.org.uk/information/factsheets/sensory_ip.html

This next site comes from The North Carolina State University: College of Urban Design, but here you are given information on how to incorporate sensory features into an accessible garden.

www.design.ncsu.edu/cud/nightsky/sensory/features.html

THERAPEUTIC GARDEN PROGRAM CURRICULUM

Here are two of our favorite books on using gardens as part of the therapeutic treatment intervention. Bibby Moore's book may be outdated, but her topics and organization of activities are still very useful.

Moore, Bibby (1989). *Growing with Gardening: A Twelve-Month Guide for Therapy, Recreation, and Education.* Chapel Hill, NC: The University of North Carolina Press.

This is the most recent book by the folks at the Chicago Botanic garden. Maria Gabaldo has a background in occupational therapy and highlights the essentials to programming with treatment in mind.

Gabaldo, Maria, King, Maryellen D., and Rothert, Eugene A. (2003). *Health Through Horticulture: A Guide for Using the Outdoor Garden for Therapeutic Outcomes.* Glencoe, IL: Chicago Botanic Garden.

TOOLS

Many manufacturers sell tools that are adaptable for all situations and needs. The AHTA Web site has a complete listing. Here are a couple of them:

Gardener Supply Company
www.gardeners.com

Access to Recreation, Inc.
www.accesstr.com

UNIVERSAL DESIGN

This Web site will help you find the information needed about universal design and how it can be integrated into your facility and landscape.

The North Carolina State University: College of Urban Design
www.design.ncsu.edu:8120/cud/

This book is an excellent resource when you are developing a garden and need advice on accessible design elements.

Rothert, Eugene (1994). *The Enabling Garden: Creating Barrier-Free Gardens.* Dallas: Taylor Publishing Co.

Index

Page numbers followed by the letter "f" indicate figures; those followed by the letter "i" indicate illustrations.

Abies balsamea (balsam fir), 19
Accessible
    garden design, 6, 13-15, 23f, 24f, 25f, 86
    garden path, 6, 13, 14-15, 17, 23f, 24f, 25f
    seating ledges, 15, 26f
    tools, 15, 85-86
    wheelchair access, 6, 13, 14-15, 17, 23f, 24f
Accessible containers
    boxes and pots, 15-16, 24f
    deep boxes, barrels, and tubs, 16, 24f, 26f
    hanging baskets, 16
    raised beds, 15, 17-18, 24f
    table planters, 16
Activities. *See also* Evaluation; Orientation
    brain and senses, 37, 39-41
    general program planning, 6, 9-12
    getting acquainted, 31-33, 35, 36
    postprogram, 76
    sight and the eye, 55, 56-57, 60
    sound and the ear, 45, 49, 52-53
    taste and smell, 68, 70-71
    touch, 61, 62-63, 67
Aging-related limitations, 28
Alzheimer, Alois, 30
Alzheimer's disease, 28, 30-31, 39
Amur chokecherry (*Prunus maackii*), 20
Annuals. *See* Flowers, annuals
Ants. *See* Bees

Anvil (incus). *See* Ear; Middle ear
Apple (*Malus* spp.), 19
Artemisia schmidtiana 'Silver Mound' (Silver mound artemesia), 19
Auditory canal. *See* Ear
Auditory nerves. *See* Ear
Auricle (outer ear, pinnea). *See* Ear

Balsam fir (*Abies balsamea*), 19
Bees
    ants, 45, 46
    bumble bees, 20, 45, 46
    honeybees, 20, 45, 46-49, 52-53, 82
    hornets, 45, 46
    mason bees, 20, 45
    wasps, 45, 46
    yellow jackets, 45, 46
Betula nigra (river birch), 20
Blood vessels. *See* Skin
Brain
    brainstem, 42, 44i
    cerebellum, 42, 44i
    cerebrum, 42, 44i
    explicit system, 39-40, 41
    hippocampus, 38
    hypothalamus, 42, 43, 44i
    implicit system, 39-40, 41
    limbic system, 38, 39-40, 41
    neurotransmitters, 38
    thalamus, 42-43, 44i

doi:10.1300/5531_07

Brainstem. *See* Brain
Bulbs. *See* Flowers, bulbs
Bumble bees. *See* Bees
Bur oak (*Quercus macrocarpa*), 20

Cerebellum. *See* Brain
Cerebrum. *See* Brain
Chocolate cosmos (*Cosmos
    atrosanguineus*), 18
Cilia. *See* Nose
Clotilde Irvine Sensory Garden and
    Program Center at the
    Minnesota Landscape
    Arboretum, 21, 22f-26f, 28
Cochlea. *See* Ear
Community recreation program, 28, 30
Convallaria majalis (lily of the valley),
    19
Cornea. *See* Eye
Cosmos atrosanguineus (chocolate
    cosmos), 18

Dadant and Sons Beekeeping
    Company, 47, 82
Dermis. *See* Skin
Drones. *See* Honeybee hive members
Dusty miller (*Senecio cineraria*), 19

Ear
    anvil, 50, 54i
    auditory canal, 50, 54i
    auditory nerves, 50, 54i
    auricle, 50
    cochlea, 50, 51, 54i
    eustachian tube, 50, 54i
    hammer, 50, 54i
    incus, 50
    inner ear, 46, 50
    malleus, 50
    middle ear, 50, 51
    outer ear, 46, 50-51, 54i
    pinnea, 50

Ear *(continued)*
    semicircular canals, 51, 54i
    stirrup bone, 46
    tympanic membrane, 50, 51, 54i
Epidermis. *See* Skin
Eustachian tube. *See* Ear
Evaluation
    postprogram, 77-79
    weekly, 33-34, 41, 49, 57, 64, 71
Evergreens. *See* Trees
Excelsior Community Recreation
    Center. *See* Community
    recreation program
Explicit system. *See* Brain
Eye
    cornea, 55, 57, 58, 59
    iris, 57, 58, 59
    lens, 55, 58, 59
    optic nerve, 58, 59
    pupil, 55, 57, 58, 59
    retina, 55, 58, 59
    sclera, 57, 58, 59
    vitreous humor, 55, 58, 59

Flavors, 70
Flowers, annuals
    chocolate cosmos (*Cosmos
        atrosanguineus*), 18
    dusty miller (*Senecio cineraria*), 19
    heliotrope (*Heliotropium
        arborescens*), 18
    marigold (*Tagetes* spp.), 18
    pansy (*Viola wittrockaina*), 19
    rose-scented geranium
        (*Pelargonium graveolens*), 19
    sweet alyssum (*Lobularia
        maritima*), 19
Flowers, bulbs
    hyacinth (*Hyacinthus orientalis*), 19
    lily (*Lilium* spp.), 19
Flowers, perennials
    lamb's ears (*Stachys byzantina*), 19
    lily of the valley (*Convallaria
        majalis*), 19

Flowers, perennials *(continued)*
  peony (*Paeonia lactiflora*), 19
  silver mound artemisia (*Artemisia schmidtiana* 'Silver Mound'), 19
Fountain grass (*Pennisetum* spp.), 19

Grasses, 19, 20
Gray Panthers, 1, 84

Hair. *See* Skin
Hair follicles. *See* Skin
Hammer (malleus). *See* Ear; Middle ear
Heliotrope (*Heliotropium arborescens*), 18
Heliotropium arborescens (heliotrope), 18
Herbs, 18, 20-21, 68, 70, 71
Hippocampus (limbic system). *See* Brain; Memory
Honeybee hive members. *See also* Bees
  drones, 47
  queen, 47, 48, 52, 53
  workers, 47, 48, 52, 53
"Honey crop" stomach, 48
Hornets. *See* Bees
Hyacinth (*Hyacinthus orientalis*), 19
Hyacinthus orientalis (hyacinth), 19
Hymenoptera, 46. *See also* Bees
Hypothalamus. *See* Brain

Implicit system. *See* Brain
Incus (anvil). *See* Ear; Middle ear
Inner ear. *See* Ear
Insects, 46. *See also* Bees
Ipomoea alba (moonflower), 19
Iris. *See* Eye

Kuhn, Maggie, 1

Lamb's ears (*Stachys byzantina*), 19
Lathyrus odoratus (sweet pea), 19
Lens. *See* Eye
Lilac (*Syringa vulgaris*), 19
Lilium spp. (lily), 19
Lily (*Lilium* spp.), 19
Lily of the valley (*Convallaria majalis*), 19
Limbic system (hippocampus). *See* Brain; Memory
Littleleaf linden (*Tilia cordata*), 19
Lobularia maritima (sweet alyssum), 19
Lonicera sempervirens (trumpet honeysuckle), 19

Malleus (hammer). *See* Ear; Middle ear
Malus spp. (apple), 19
Marigold (*Tagetes* spp.), 18
Mason bees. *See* Bees
Memory, 38, 40. *See also* Brain
Microvilli. *See* Papillae; Tongue
Middle ear. *See also* Ear
  anvil (incus), 50, 54i
  hammer (malleus), 50, 54i
  stirrup bone, 46
  tympanic membrane, 50, 51, 54i
Minnesota Landscape Arboretum, 5, 82. *See also* Clotilde Irvine Sensory Garden and Program Center at the Minnesota Landscape Arboretum
Miscanthus (grass), 20
Moonflower (*Ipomoea alba*), 19

Nerve endings. *See* Skin
Neurotransmitters. *See* Brain
Nose
  cilia, 69, 72, 73, 75
  olfactory bulb, 69, 70, 73, 75
  olfactory epithelium, 72, 73, 75
  olfactory nerves, 69, 72, 73, 75

Olfactory bulb. *See* Nose
Olfactory epithelium. *See* Nose
Olfactory nerves. *See* Nose
Optic nerve. *See* Eye
Orientation
  children, 30-31
  elders, 28-29
Outer ear (auricle, pinnea). *See* Ear

Paeonia lactiflora (peony), 19
Panicum (grass), 20
Pansy (*Viola wittrockaina*), 19
Papillae. *See also* Tongue
  filiform, 71
  fungiform, 71
  microvilli, 72
  taste buds, 69, 70, 72, 74
  vallate, 71-72
Parkinson, James, 31
Parkinson's disease, 28, 30, 31
Pelargonium graveolens (rose-scented
  geranium), 19
Pennisetum spp. (fountain grass), 19
Peony (*Paeonia lactiflora*), 19
Perennials. *See* Flowers, perennials
Pheromone, 48
Pinnea (auricle, outer ear). *See* Ear
Pinus strobus (white pine), 20
Pores. *See* Skin
Postevaluation. *See* Evaluation
Prunus maackii (amur chokecherry), 20
Pupil. *See* Eye

Quaking aspen, 20
Queen. *See* Honeybee hive members
Quercus macrocarpa (bur oak), 20

Raised bed planter construction, 17-18.
    See also Accessible containers
Retina. *See* Eye
River birch (*Betula nigra*), 20
Rosa spp. (rose), 19

Rose (*Rosa* spp.), 19
Rose-scented geranium (*Pelargonium
  graveolens*), 19

Saliva. *See* Tongue
Salivary glands. *See* Tongue
Sclera. *See* Eye
Sebaceous (oil) glands. *See* Skin
Sebum. *See* Skin
Semicircular canals. *See* Ear
Senecio cineraria (dusty miller), 19
Senses
  sight, 55-60
  smell, 68-75
  sound, 45-54
  taste, 68-75
  touch, 61-67
Sensory garden
  design, 21, 22f-26f, 84-85
  garden sounds, 14, 20
  plants, 18-21, 24f, 25f, 26f
Shrubs
  lilac (*Syringa vulgaris*), 19
  rose (*Rosa* spp.), 19
Sight. *See* Senses
Silver mound artemisia (*Artemesia
  schmidtiana* 'Silver Mound'),
  19
Skin
  blood vessels, 62, 64, 66
  dermis, 62, 64, 65, 66
  epidermis, 62, 64, 65, 66
  hair, 64, 66
  hair follicles, 62, 64-65, 66
  nerve endings, 62, 64, 65, 66
  pores, 65, 66
  sebaceous (oil) glands, 64, 65, 66
  sebum, 64, 65
  subcutaneous layer, 64, 65, 66
  sweat glands, 64, 65, 66
Smell. *See* Senses
Sojourn Adult Day Care Center, 28
Sound. *See* Senses
Stachys byzantina (lamb's ears), 19

Stirrup bone. *See* Ear; Middle ear
Strategies, 9-12, 77-79
Strokes, 28, 30, 31
Subcutaneous layer. *See* Skin
Sweat glands. *See* Skin
Sweet alyssum (*Lobularia maritima*), 19
Sweet pea (*Lathyrus odoratus*), 19
Syringa vulgaris (lilac), 19

Tagetes spp. (marigold), 18
Taste. *See* Senses
Taste buds. *See* Tongue
Thalamus. *See* Brain
Tilia cordata (littleleaf linden), 19
Tongue
    microvilli, 72
    papillae, 70, 71-72, 74
    saliva, 72, 74
    salivary glands, 72, 74
    taste buds, 69, 70, 72, 74
Touch. *See* Senses
Trees
    amur chokecherry (*Prunus maackii*), 20
    apple (*Malus* spp.), 19
    balsam fir (*Abies balsamea*), 19

Trees *(continued)*
    bur oak (*Quercus macrocarpa*), 20
    littleleaf linden (*Tilia cordata*), 19
    quaking aspen (*Populus tremuloides*), 20
    river birch (*Betula nigra*), 20
    white pine (*Pinus strobus*), 20
Trumpet honeysuckle (*Lonicera sempervirens*), 19
Tympanic membrane. *See* Ear; Middle ear

Vines
    moonflower (*Ipomoea alba*), 19
    sweet pea (*Lathyrus odoratus*), 19
    trumpet honeysuckle (*Lonicera sempervirens*), 19
Viola wittrockaina (pansy), 19
Vitreous humor. *See* Eye

Wasps. *See* Bees
White pine (*Pinus strobus*), 20
Workers. *See* Honeybee hive members

Yellow jackets. *See* Bees

Milton Keynes UK
Ingram Content Group UK Ltd.
UKHW022048141024
449569UK00031B/1550